Return-to-Play after Lower Limb Muscle Injury in Football

Gian Nicola Bisciotti • Alessandro Corsini
Piero Volpi

Return-to-Play after Lower Limb Muscle Injury in Football

The Italian Consensus Conference Guidelines

Springer

Gian Nicola Bisciotti
Qatar Orthopaedic and Sports Medicine Hospital
Doha
Qatar

Alessandro Corsini
F.C. Internazionale
Milan
Italy

Piero Volpi
Humanitas Clinical and Research Center
IRCCS.Rozzano
Head of Medical Staff of
F.C. Internazionale Football Club
Milan
Italy

ISBN 978-3-030-84949-8 ISBN 978-3-030-84950-4 (eBook)
https://doi.org/10.1007/978-3-030-84950-4

This Springer imprint is published by the registered company Springer Nature Switzerland AG
The registered company address is: Gewerbestrasse 11, 6330 Cham, Switzerland

Introduction

The "Italian Consensus Conference on Guidelines on return to play after lower limb muscle injury in football" was organized by the Italian Society of Arthroscopy in Milan, on 31 August 2018, with the participation of 66 national and international experts with different medical backgrounds, i.e., orthopedic surgeons (19), sports physicians (7), radiologists (5), rehabilitation physicians (3), sport physiologists (2), general surgeons (2), family physicians (2), physiotherapists (10), physical trainers (15), and psychologist (1).

The selection of the Consensus Conference participants was based on their Hirsch index, the number of publications concerning muscle injuries, and experience in the clinical evaluation, medical treatment, and rehabilitation of MI. The experts did not represent any organizations. All experts who participated, directly or indirectly, in the Consensus Conference must be considered as coauthors of this book. Indeed, without their contribution and participation this book would never have been realized. So we want to thank them sincerely and to remember all their names:

Giampietro Alberti, Alessandro Aprato, Matteo Artina, Alessio Auci, Corrado Bait, Andrea Belli, Giuseppe Bellistri, Pierfrancesco Bettinsoli, Alessandro Bisciotti, Andrea Bisciotti, Stefano Bona, Marco Bresciani, Andrea Bruzzone, Roberto Buda, Michele Buffoli, Matteo Callini, Gianluigi Canata, Davide Cardinali, Gabriella Cassaghi, Lara Castagnetti, Sebastiano Clerici, Barbara Corradini, Cristina D'Agostino, Enrico Dellasette, Francesco Di Pietto, Drapchind Enrica, Cristiano Eirale, Andrea Foglia, Francesco Franceschi, Antonio Frizziero, Alberto Galbiati, Carlo Giammatei, Philippe Landreau, Claudio Mazzola, Biagio Moretti, Marcello Muratore, Gianni Nanni, Roberto Niccolai, Claudio Orizio, Andrea Pantalone, Federica Parra, Giulio Pasta, Paolo Patroni, Davide Pelella, Luca Pulici, Alessandro Quaglia, Stefano Respizzi, Luca Ricciotti, Arianna Rispoli, Francesco Rosa, Alberto Rossato, Italo Sannicandro, Claudio Sprenger, Chiara Tarantola, Fabio Gianpaolo Tenconi, Giuseppe Tognini, Fabio Tosi, Giovanni Felice Trinchese, Paola Vago, Marcello Zappia, Zarko Vuckovich, Raul Zini, Michele Trainini, Karim Chamari.

The "Italian Consensus Conference on Guidelines on return to play after lower limb muscle injury in football" is part of a project that includes a triptych of three "Italian Consensus Conferences" with the other two entitled respectively: "Groin Pain Syndrome Italian Consensus Conference on terminology, clinical evaluation and imaging assessment in groin pain in athlete" and "Italian Consensus Conference on Guidelines for Conservative Treatment on Lower Limb Injuries in Athlete." During this Consensus Conference, the invited experts discussed and approved a consensus composed of two sections.

The first section was composed of:

Terminology relating to return to play (RTP).

Return to training (RTT): decision-making process.

RTP: decision-making process.

The role of imaging: in RTT and RTP decisions-making process.

The biopsychosocial model and RTP decisions-making process.

The second section was composed of:

RTT and RTP following hamstring injuries.

RTT and RTP following quadriceps injuries.

RTT and RTP following adductor injuries.

RTT and RTP following soleus-gastrocnemius injuries.

RTT and RTP following short external hip rotator injuries.

RTT and RTP following iliopsoas injuries.

This book represents the summary of the Consensus Conference works and hopes to provide a valuable help to sports physicians interested in this specific area.

Paris

April 30th, 2021

Gian Nicola Bisciotti
Alessandro Corsini
Piero Volpi

Contents

1 The Italian Consensus Conference on Return to Play After Lower Limb Muscle Injury in Football

1.1 Introduction

What is a "Consensus Conference" (CC)? The CC is a methodology for answering the questions related to efficacy, risks, and clinical applications of biomedical or public health interventions. The subjects discussed during a CC represent specific controversies in clinical practice. Furthermore, the aim of a CC is also to orient future research. A CC is carried out through the production of evaluation reports of the scientific literature, discussed by a "jury" made up of health professionals and other professional and social figures. The National Institutes of Health (NIH) developed the CC model in the United States in the 1970s, as a method for addressing complex problems relating to health interventions and for orienting research. In Europe the first CCs were organized in Denmark in the 1980s by the Danish Board of Technology (Danish Commission on Technology). The CC aims at defining the "state of the art" in relationship to a specific care problem, through an explicit process in which scientific information is evaluated and discussed by a "jury." Such jury is composed not only of health professionals but also of other professional and social members. In recent years CCs have been enriched by an intense specific preparatory activity, which has been entrusted to working groups in charge of preparing in-depth documents. This has become necessary to overcome the limits and operational difficulties, to concentrate the discussion on all issues in a single day, and to find a consensus from the jury. The documents produced by each group are provided to the members of the jury well in advance before the CC and will form the basis for the discussion. Conversational dynamics is a very important aspect in successful consensus conferences. Some jury members may tend to dominate the conversation. It is a drawback that can be avoided by introducing a (well-trained) facilitator, hence the issue of how can facilitators be assessed. Paradoxically, a CC may even have a negative effect (opposite to its goal). Indeed, a CC could lead the jury members to move to the extreme in their opinions. In other words, the jury members essentially rally around their own view in the presence of opposing views. Another paradox of CCs is their power to mask differences in opinions. The goal of consensus conferences is for the jury members to deliberate and reach a consensus over a particular issue. However, this need to reach a consensus can have the unintended side effect of masking differences in opinion, particularly if some jury members are less outspoken during the discussion. In any case, despite these possible negative aspects, the CC represents a formidable means for the improvement of scientific knowledge.

G. N. Bisciotti et al., *Return-to-Play after Lower Limb Muscle Injury in Football*,
https://doi.org/10.1007/978-3-030-84950-4_1

1.2 The Italian Consensus Conference on Return to Play After Lower Limb Muscle Injury in Football

The Italian Consensus Conference on return to play after lower limb muscle injury in football (Bisciotti et al. 2019), held in Milan on August 31, 2018, was aimed at making the concepts of return to training (RTT) and return to play (RTP) clearer and more objective. Furthermore, the CC tried to identify some weak points concerning both RTT and RTP which may seriously jeopardize their reliability. With this aim in mind, the Italian Society of Arthroscopy organized this Conference in Milan, on August 31, 2018, which saw the participation of 66 national and international experts across several disciplines: orthopedic surgeons (19), sports physicians (7), radiologists (5), rehabilitation physicians (3), sports physiologists (2), general surgeons (2), family physicians (2), physiotherapists (10), physical trainers (15), and psychologists (1).

The selection of experts was based upon their Hirsch index and/or the number of publications concerning muscle injuries in football and/or their experience in the clinical evaluation, medical treatment, and rehabilitation of muscle injuries in football players. The experts did not represent any commercial organizations at the time of the consensus meeting.

Prior to the CC, two senior authors (GNB and PV) performed a narrative review of the literature regarding RTP decision-making process, in sporting activities in general and in football specifically. The review process was conducted as follows:

1. An independent search was performed by both authors, with no language limitation applied.
2. Databases searched were MEDLINE, EMBASE, EXCERPTA MEDICA, Cochrane Central Register of Controlled Trials, and the Cochrane Database of Systematic Review. Gray literature (i.e., conferences, abstracts, thesis, and unpublished reports) was not considered.

Table 1.1 Inclusion and exclusion criteria

Inclusion criteria: *P(atient and problem)*: RCT, case series studies, and consensus statement investigating lower muscle injuries in sport population. *I(ntervention)*: Conservative treatment of lower muscle injuries. *C(omparison)*: Comparison between different types of muscle injuries classification and different types of conservative treatments. *O(utcome)*: Outcome in terms of time loss injury, level of outcome, level of return to play, complication, and sequelae
Exclusion criteria: *P(atient and problem)*: RCT, case series studies, and consensus statement investigating lower muscle injuries in a non-sport population. *I (ntervention)*: Surgical treatment of lower muscle injuries. *C(omparison)*: Comparison between conservative and surgical treatments. *O(utcome)*: Not specified outcome in terms of time loss injury, level of outcome, level of return to play, complication, and sequelae

3. Studies that did not meet our inclusion criteria were excluded. The inclusion and exclusion criteria, according to PICO method (Cooke et al. 2012), are shown in Table 1.1.

After the review, the authors produced a comprehensive summary document made up of two distinct sections: (1) RTP decision-making process's general principles and (2) return to training and RTP decision-making process following lower limb muscle injuries. The document was presented to each expert a week ahead of the CC and was considered as the starting point for the discussion. The two senior authors had the role of facilitator (GNB) and chairman (PV) during the CC.

1.3 Consensus Conference Proceedings

The CC was composed of two distinct sections.

The discussion on RTP decision-making process's general principles (Sect. 1.1) was in turn composed of five sub-sections:

1. Appropriateness of the term RTP
2. The return to training decision-making process (RTT decision-making process)
3. The return to play decision-making process (RTP decision-making process)

4. The role of imaging in RTT decision-making and RTP decision-making processes
5. The biopsychosocial model

In the CC, the term "decision-making process" (DMP) refers to the evidence-based criteria utilized for decision on RTT and RTP (Van der Horst et al. 2016, 2017).

The second section was focused on RTT and RTP-DMP following lower limb muscle injuries in football. This discussion addressing RTP-DMP for the lower limb muscle groups involved in football muscle injuries is composed of four sub-sections:

1. RTT and RTP-DMP following hamstring injuries
2. RTT and RTP-DMP following quadriceps injuries
3. RTT and RTP-DMP following adductors injuries
4. RTT and RTP-DMP following soleus-gastrocnemius injuries

Following a Delphi procedure, each document was presented by the facilitator (GNB), a plenary discussion conducted by the chairman (PV) followed, and finally it was approved via a voting process.

The CC participants voted for each document, using a Likert scale of 0–10, where 0 reflected complete disagreement, 5 neither agreement nor disagreement, and 10 complete agreement. The discussions continued until a mean score of >7.5 was reached (Griffin et al. 2016; Bisciotti et al. 2016, 2018; Vanbelle and Lesaffre 2017), and the voting process enabled the chairman to interrupt the discussion if, in his opinion, a final decision could not be reached. The first section document required five voting rounds (for five separate discussions), while the second section document required four voting rounds (for four separate discussions) to reach consensus. Amendments were made after each voting round following discussion among the CC group. Consensus was reached in all cases after each discussion phase (i.e., for each voting round the mean score of >7.5 was reached). The voting results are shown in Tables 1.2 and 1.3.

At the end of the CC, consensus was reached on the following: definition of the terms RTT and RTP in football, the criteria for RTT and RTP in football, the appropriate use of clinical and imaging assessment, laboratory and field tests for RTT following lower limb muscle injury, and the identification of return to play objective criteria following lower limb muscle injury were discussed and approved (level of evidence IV, grade of recommendation D). This book illustrates in detail and thoroughly all the topics discussed and approved during the CC.

Table 1.3 The results of the different voting rounds concerning document 2 (return to training and RTP-DMP following lower limb muscle injuries in football)

	Voting 1	Voting 2	Voting 3	Voting 4
Average score	9.24	9.64	9.54	9.72
Standard deviation (±)	0.49	0.39	0.43	0.35

Table 1.2 The results of the different voting rounds concerning document 1 (RTP-DMP general principles)

	Voting 1	Voting 2	Voting 3	Voting 4	Voting 5
Average score	9.76	9.76	9.80	9.72	9.96
Standard deviation (±)	0.33	0.33	0.30	0.35	0.32

References

Bisciotti GN, Volpi P, Zini R, Auci A, Aprato A, Belli A, Bellistri G, Benelli P, Bona S, Bonaiuti D, Carimati G, Canata GL, Cassaghi G, Cerulli S, Delle Rose G, Di Benedetto P, Di Marzo F, Di Pietto F, Felicioni L, Ferrario L, Foglia A, Galli M, Gervasi E, Gia L, Giammattei C, Guglielmi A, Marioni A, Moretti B, Niccolai R, Orgiani N, Pantalone A, Parra F, Quaglia A, Respizzi F, Ricciotti L, Pereira Ruiz MT, Russo A, Sebastiani E, Tancredi G, Tosi F, Vuckovic Z. Groin Pain Syndrome Italian Consensus Conference on terminology, clinical evaluation and imaging assessment in groin pain in athlete. BMJ Open Sport Exerc Med. 2016;2(1):e000142.

Bisciotti GN, Volpi P, Amato M, Alberti G, Allegra F, Aprato A, Artina M, Auci A, Bait C, Bastieri GM, Balzarini L, Belli A, Bellini G, Bettinsoli P, Bisciotti A, Bisciotti A, Bona S, Brambilla L, Bresciani M, Buffoli M, Calanna F, Canata GL, Cardinali D, Carimati G, Cassaghi G, Cautero E, Cena E, Corradini B, Corsini A, D'Agostino C, De Donato M, Delle Rose G, Di Marzo F, Di Pietto F, Enrica D, Eirale C, Febbrari L, Ferrua P, Foglia A, Galbiati A, Gheza A, Giammattei C, Masia F, Melegati G, Moretti B, Moretti L, Niccolai R, Orgiani A, Orizio C, Pantalone A, Parra F, Patroni P, Pereira Ruiz MT, Perri M, Petrillo S, Pulici L, Quaglia A, Ricciotti L, Rosa F, Sasso N, Sprenger C, Tarantola C, Tenconi FG, Tosi F, Trainini M, Tucciarone A, Yekdah A, Vuckovic Z, Zini R, Chamari K. Italian consensus conference on guidelines for conservative treatment on lower limb muscle injuries in athlete. BMJ Open Sport Exerc Med. 2018;4(1):e000323.

Bisciotti GN, Volpi P, Alberti G, Aprato A, Artina M, Auci A, Bait C, Belli A, Bellistri G, Bettinsoli PF, Bisciotti AL, Bisciotti AN, Bona S, Bresciani M, Bruzzone A, Buda R, Buffoli M, Callini M, Canata GL, Cardinali D, Cassaghi G, Castagnetti L, Clerici S, Corradini B, Corsini A, D'Agostino C, Dellasette E, Di Pietto F, Drapchind E, Eirale C, Foglia A, Franceschi F, Frizziero A, Galbiati A, Giammattei C, Landreau P, Mazzola C, Moretti B, Muratore M, Nanni G, Niccolai R, Orizio C, Pantalone A, Parra F, Pasta G, Patroni P, Pellela D, Pulici L, Quaglia A, Respizzi S, Ricciotti L, Rispoli A, Rosa F, Rossato A, Sannicandro I, Sprenger C, Tarantola C, Tenconi FG, Tognini G, Tosi F, Trinchese GF, Vago P, Zappia M, Vuckovic Z, Zini R, Chamari K. Italian Consensus Conference on return-to-play after lower limb muscle injury in football (Soccer). BMJ Open Sport Exerc Med. 2019;5(1):e000505.

Cooke A, Smith D, Booth A. Beyond PICO the SPIDER tool for qualitative evidence synthesis. Qual Health Res. 2012;22(10):1435–43.

Griffin DR, Dickenson EJ, O'Donnell J, et al. The Warwick Agreement on femoroacetabular impingement syndrome (FAI syndrome): an international consensus statement. Br J Sports Med. 2016;50:1169–76.

Van der Horst N, van de Hoef S, Reurink G, et al. Return to play after hamstring injuries: a qualitative systematic review of definitions and criteria. Sports Med. 2016;46(6):899–912.

Van der Horst N, Backx F, Goedhart EA, Huisstede BM, HIPS-Delphi Group. Return to play after hamstring injuries in football (soccer): a worldwide Delphi procedure regarding definition, medical criteria and decision-making. Br J Sports Med. 2017;51(22):1583–91.

Vanbelle S, Lesaffre E. Modeling agreement on bounded scales. Stat Methods Med Res. 2017;1:096228021770570.

General Principles for Return to Training and Return to Play

2

2.1 The Appropriateness of the RTP Term

The first issue to be addressed is whether the term RTP is appropriate (Ardern et al. 2016c; Bisciotti et al. 2018; Bisciotti and Volpi 2018).

The RTP was defined by the consensus statement developed by The American Academy of Orthopaedic Surgeons and American College of Sports Medicine as follows:

> The decision-making process of returning an injured or ill athlete to practice or competition. This ultimately leads to medical clearance of an athlete for full participation in sports (Herring et al. 2002, 2012, 2013).

Questioning the correctness of this term is not related to semantic reasons but exclusively to a principle of correct application. Indeed, the term "return to play" in football implicitly means a return to "full availability of play." In other words, if a player meets the requirements adopted for the RTP decision-making process (DMP), let's say on Saturday morning, then he could theoretically be available for the competition on the next day. Obviously, this is a clear extremization which, however, is useful in clarifying the concept. In the majority of cases, in fact, the RTP involves a period—whose length depends on both the type of injury suffered and the time-loss injury sustained—of progressive reinsertion in the team. As it is easy to understand, we are far from being able to identify the RTP "tout court" with the immediate and full availability of the player, namely, "availability without restrictions" (Ardern et al. 2016c; Bisciotti et al. 2018). Indeed, the RTP is preceded by a period, be it shorter or longer, in which the player is reinserted with the team but is subject to some limitations. Such limitations can concern many aspects of training: its intensity, its duration, or some movements and situations that are still particularly at risk for reinjury.

For these reasons, the CC considered that it is necessary to introduce a clarification regarding the term RTP in accordance with the 2016 consensus statement on the return to sport (Ardern et al. 2016c).

The CC agreed to adopt the term RTT for the situation in which the player is reinserted in the team training but with some possible limitations. On the contrary, the CC agreed that the term RTP means the situation in which the player once again reaches full availability both for training without any restriction and for competition. In other words, RTT is linked to the concept of "returning to training with possible restrictions," while RTP is connected to the concept of "returning to training and competition without any kind of restriction." In this regard it is necessary to underline that RTT and RTP are based on different decision criteria. Indeed, while RTT is inspired by clinical-functional decision-making criteria, RTP is based on functional performance criteria, thus passing from a purely medical evaluation to a more technical-athletic one.

G. N. Bisciotti et al., *Return-to-Play after Lower Limb Muscle Injury in Football*,
https://doi.org/10.1007/978-3-030-84950-4_2

2.2 The RTT Basic Principles

The RTP-DMP must be based on clinical criteria, imaging criteria, functional tests, and functional "injury-type-dependent" tests.

In other words, for each type of injury, the following points must be identified and respected:

1. Identification of a series of specific clinical tests for the type of considered injury
2. Identification of a battery of "laboratory" tests (LT) specific for the functional deficit that the type of considered injury may cause
3. Identification of a battery of "field" tests (FT) specific for the functional deficit that the type of considered injury may cause

Points *i* and *ii* must be analyzed through a thoughtful judgment based on the clinical evaluation and analysis of the imaging, taking into account the performance model, the role, and the functional requirements of the athlete.

2.3 The RTP Basic Principles

The evaluation of RTP is substantially a judgment of suitability for the resumption of full training load without restrictions, as well as of the ability to conveniently face the competition.

The DMP for RTP, which follows that of RTT, is an evaluation that is based more on a judgment of "functionality" and "performance capacity" rather than "clinical-functional suitability" like for RTT.

The RTP-DMP makes use of the global positioning system (GPS) technology, which allows to collect and analyze the indices on which the RTP is based. The details of RTP-DMP will be discussed in Chap. 4.

2.4 The Evaluation of Physical Fitness

Numerous studies have shown the association of decreased aerobic fitness with the risk of incurring muscle injuries (Hopper et al. 1995; Chomiak et al. 2000; Ostenberg and Roos 2000; Bell et al. 2000; Dvorak et al. 2000; Murphy et al. 2003). Other studies show that injuries requiring long rehabilitation periods, during which mainly low-intensity physical activities are carried out, decrease aerobic fitness (Almeida et al. 2018). Indeed, already only 20 days of inactivity lead to a significant decrease in VO_2max values (Convertino 1997; Bringard et al. 2010). Therefore, in rehabilitation plans entailing more than 20 days of low-level aerobic activity, it is advisable to check the value of VO_2max and/or the corresponding maximal aerobic speed (MAS) through an incremental speed running test. The CC experts suggest that the RTP-DMP be corroborated by means of a valid test for determining VO_2max (Heck et al. 1985; Chamari et al. 2004, 2005; Badawy and Muaidi 2018) in which the player has recorded a value of VO_2max, and consequently of MAS, equal to at least 90% of the pre-injury value.

2.5 Evidence on RTT and RTP

To date, the adoptable criteria for RTT and RTP cannot be based on solid evidence, given the widespread lack of consensus on this specific topic. Indeed, there are still many obscure points, such as the real incidence, and the consequent influence, of the reinjuries on the DMP for RTT and RTP. Another point to clarify is the precise percentage of athletes who return to the pre-injury performance level. Furthermore, there is a total lack of consensus concerning the tests both for RTT and in RTP (Ardern et al. 2016c). However, what stands out is that a correct initial clinical evaluation is associated with a good outcome (Weiler 2016). For all the reasons above, it is clear that, at the current state of knowledge, the RTT-DMP and RTP-DMP must necessarily be based on expert advice and therefore on a rather low level of evidence (level of evidence IV—grade of recommendation D).

2.6 Risk in the Short, Medium, and Long Term

Based on current literature, a further and important point is represented by DMP concerning the RTT, and consequently the RTP, only taking into consideration the so-called "short-term risks" (Shrier 2015). It is a serious limitation for the decision-making reliability in the medium to long term. This limitation becomes particularly evident for some clinical frameworks often involving possible sequelae in the medium/long term. It is, for example, the case of the ACL reconstruction, of the different techniques of cartilage repair and/or regeneration, or of the massive injuries of some muscle groups, such as the hamstring. Thus, it is strongly advised that the DMP be not based solely on criteria that calculate only the short-term risk, but rather on a clinical-functional reasoning that considers all—or at least most of—the medium–/long-term risk factors able to put at risk the continuation of the athlete's career (or his health in a general sense) in the medium-long term. A suitable example could be RTP in case of cartilage repair and/or regeneration surgical intervention. In this case, only considering the short-term risks requires the calculation of the risk of injury recurrence alone. On the contrary, a medium-/long-term DMP also considers the risk of possible early prosthesization and its consequences on the subject's future quality of life. The principle of medium-/long-term decision-making evaluation is dictated not only by professional ethics (which goes beyond the mere assessment of the player's ready use) but also by the not undervaluable fact that the player represents, for the Club to which he belongs, a financial investment to be optimally protected.

2.7 The Concept of "Tolerable Risk" in RTT-DMP

Another crucial point of the RTT and RTP-DMP is represented by the concept of "tolerable risk." Tolerable risk (TR) is the maximum level of risk acceptable for different short-term and long-term outcomes associated with RTT (Ardern et al. 2016a, b; Shrier 2015; Creighton et al. 2010). TR is attributed equally to medical and technical staff, team management, and the player. Therefore, it is primarily a "shared risk" which presupposes a consequent sharing of responsibilities. The concept of "shared risk" requires the identification of specific figures having decision-making/consultative power concerning RTT and RTP.

TR is not shared in the event of life-threatening situations (e.g., concussion in which the player has a reduced level of consciousness/decision-making ability). Under such circumstances, the sole and final decision on RTT and RTP depends entirely upon the medical team assessment. Apart from these particular and unavoidable eventualities, the TR represents an extremely variable concept in function of its context. Think of how greater TR may be in a cup final compared to a friendly match, hence the need to establish, in the RTT and RTP-DMP, a maximum threshold of TR (making an evaluation as objective as possible) and to compare it with the real risk (RR): if the value of RR exceeds that of TR the decision of the RTT and/or the RTP must necessarily be postponed. In the specific case of muscle and/or tendon injuries, TR corresponds to the objective quantification of the maximum mechanical load that can be tolerated by damaged muscle tissues. Here, the TR must be calculated based on:

1. Clinical examination
2. Imaging
3. Functional tests

Furthermore, TR can be influenced by several factors: whether it is an acute or overuse injury, a first-time injury or a reinjury, its degree of severity and anatomical location, its type (i.e., monoarticular muscle, biarticular muscle, myotendinous junction, in proximity to the central tendon, etc.), and other biological, endocrine-metabolic, and gender-related factors.

TR may also need to include economic evaluations; a typical example is when the player is directly involved in a market negotiation (i.e., transfer). In any case, please note that the medical staff has the responsibility to act in the best interests of the player's long-term health, regardless of any contractual negotiation. The tolerance risk flowchart is shown in Fig. 2.1.

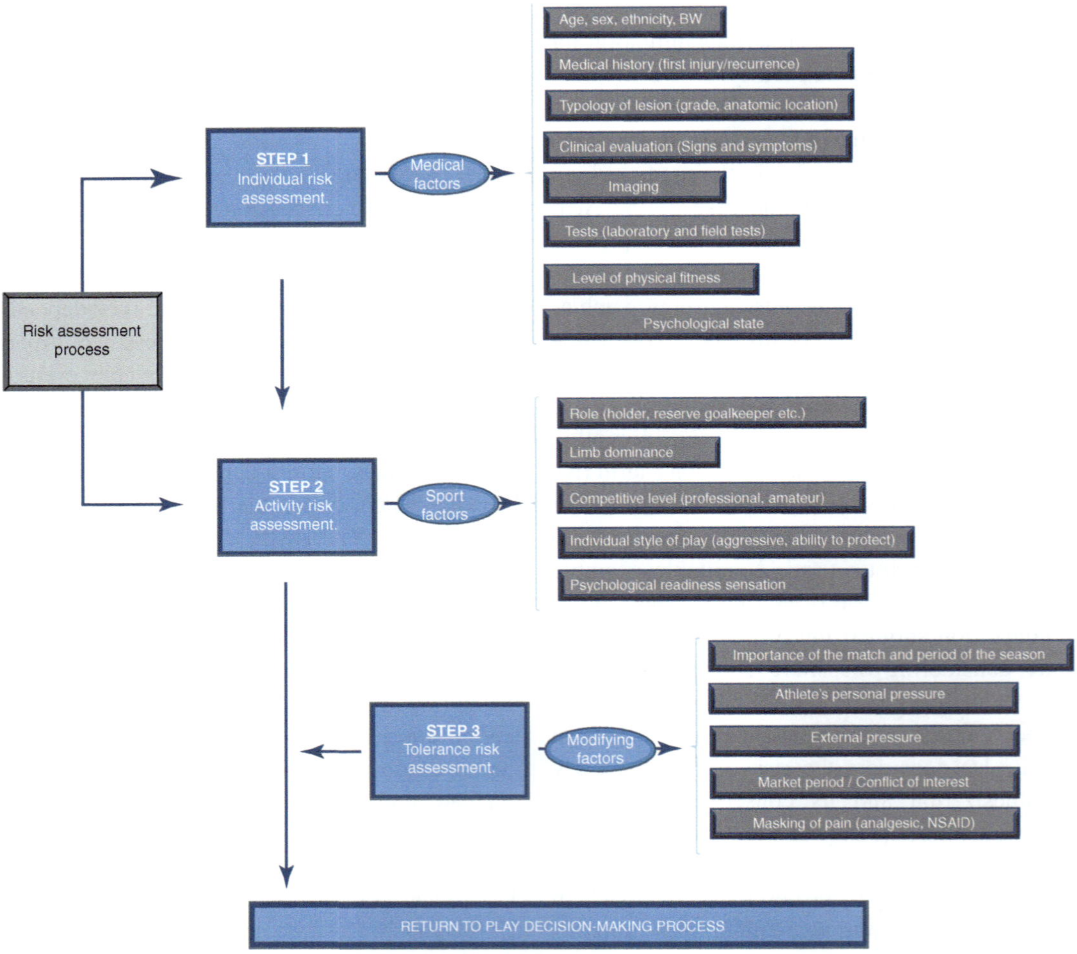

Fig. 2.1 Tolerance risk flowchart. The first step is the "individual risk assessment," and the second step is the "activity risk assessment." First and second steps together correspond to the "risk assessment process." The third step (tolerance risk assessment) affects the risk assessment process in the return to play decision-making process. Key: *BW* body weight

2.8 The "Intensity-Type-Timing" Principle

In the first RTT phase, it would be appropriate to refer to the principle of "intensity-type-timing." In accordance with this principle, the terms "intensity," "type," and "timing" must be interpreted as follows:

1. The term "intensity" refers to the intensity of the exercise required (such as the intensity required in small side games or during exercises based on changes of direction, sprints, etc.).
2. The term "timing" refers to the maximum duration of some exercises and/or situations (e.g., the request for a maximum of minutes during small side games or training games).
3. The term "type" indicates the type of exercises and/or proposed situations.

2.9 The Basic Principles of Functional Tests

Concerning the mechanical stress applied to damaged muscle tissues during functional tests, the latter must necessarily be the most similar to the one occurring in the real game situation (Creighton et al. 2010). To do so, the following parameters should be respected:

1. Forces required during muscle contraction.
2. Speed required during movement.
3. Power expressed during movement.
4. Type of movement required (i.e., specific or non-specific to the football model, for example straight line running is a nonspecific movement while cutting during a run is a specific movement).
5. Specificity of the required movement (i.e., specific or nonspecific in comparison to the movement/movements that can cause a reinjury in the previously injured muscles). For example, a sprint is a specific risk movement for biceps femoris muscle injury, and kicking is a specific risk movement for rectus femoris muscle injury.

2.10 The Objectivity of the Decision Criteria

Decisions for RTT and RTP should be based on objective criteria, which is why clinical and functional investigations should be quantified in a precise and appropriate manner. If an objective evaluation is impossible, a subjective assessment that is somewhat quantifiable, such as a reported visual analog scale (VAS) value, is also acceptable. Pain is, in any case, considered an essential parameter of evaluation in the DMP by many authors (Saal 1991; Cantu 2000; Bowen et al. 2004; Curl 2004; Drake et al. 2004; Eck and Riley 2004; Lord and Winell 2004; McCarty et al. 2004; Park et al. 2004; Dimberg and Burns 2005; Kaeding et al. 2005; Kovacic and Bergfeld 2005; Orchard et al. 2005; Burnett and Sonntag 2006; Diehl et al. 2006; Kuhn 2006; Smurawa and Congeni 2007; Krabak and Kennedy 2008; Miller et al. 2009). This widespread and shared opinion regarding the important role of pain is underpinned by the fact that the presence of a residual pain symptomatology at the level of the injured tissues is strongly suggestive of a still incomplete healing process. Furthermore, many authors strongly emphasize the idea that RTP can only be granted with full resolution of symptoms (Lord and Winell 2004; Eddy et al. 2005; Dunn et al. 2006; Elias et al. 2007; Miller et al. 2009).

2.11 The Patient Centrality

The RTT and RTP-DMP must respect the principle of patient centrality. This principle is respected provided that two conditions are met:

1. The medium- and long-term risk is also considered as well as the short-term risk.
2. The patient is seen as a "decision maker" who is active throughout the whole RTT and RTP-DMP (Fig. 2.2).

2.12 The Different Focus Types

It is important to note that often the RTT and RTP concepts are different in the eyes of the athlete, the coach, or the medical staff. An athlete may often see the DMP as aimed at minimize reducing the time needed for RTT and RTP (focus on timing), a coach (and sometimes the athlete too) will most of the time strive for the return to best competitive level (focus on the performance), and the medical staff will target DMP toward avoiding reinjuries (focus on the outcome). These three different interpretations of a concept often cause important misunderstanding. Nevertheless, focusing on the outcome must necessarily be prioritized over the other views, which must therefore be considered like subordinates to this latter.

2.13 The Decision-Making Process

The RTT and RTP-DMP should be based on a decision process which runs parallel like a continuum to the rehabilitation process. It is important to prevent the decision process from becoming isolated and detached from the rehabilitation pathway performed by the athlete. Each decision must be taken collectively by the stake-

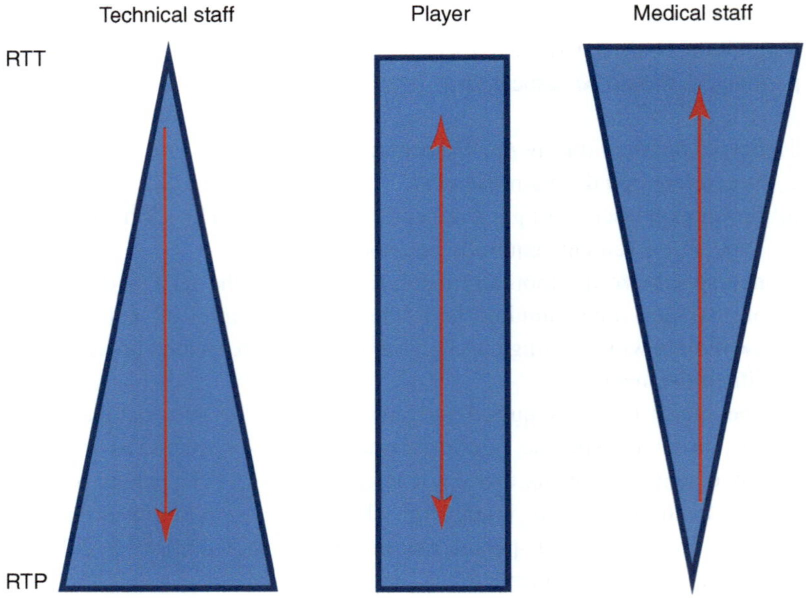

Fig. 2.2 As shown in the figure, the technical staff has a minimal incidence in the RTT-DMP, which is mostly a medical decision. On the contrary, the RTP-DMP, which is based more on a judgment of "functionality" and "performance capacity" rather than "clinical-functional suitability," is mainly affected by the technical staff judgment. It is important to note that the player is an active "decision maker" throughout the whole RTT and RTP- DMP

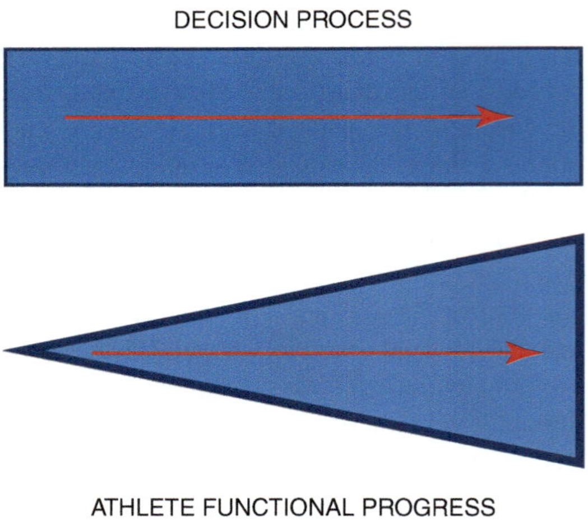

Fig. 2.3 The decision-making process must be a continuum based on the functional progress of the athlete

holders, keeping the player at the center of the DMP. Furthermore, the functional progress of the player during his rehabilitation process must be the guide for DMP (Fig. 2.3).

2.14 Closed and Open Skill Tests

During the RTT and RTP, the DMP should be based on both general and specific tests. General tests must be based on "closed skill" situations, while specific tests must be based on "open skill" situations, which should be as specific as possible in relationship to the sport activity model (in this case the football model). In other words, the specific tests must reproduce the performance model or part of it (Sheppard et al. 2006; Gabbett and Benton 2009; Lockie et al. 2014). It is important to remember that a "closed skill" situation requires the subject a motor reaction (i.e., a movement) that is already known from the beginning in relation to a known stimulus. The typical example of a "closed skill" situation is to start a sprint at the time of a previously agreed sound signal (i.e., the sprint starts from the starting blocks in athletics). On the contrary, during an "open skill" situation, the subject must react with an unpredictable movement in response to an unknown stimulus. A football game represents a typical "open skill" situation.

2.15 The Biopsychosocial Model

RTT and RTP-DMP are heavily influenced by the psychosocial context within which they occur (Wiese-Bjornstal et al. 1998; Atkins et al. 2012; Ayers et al. 2013; Pincus et al. 2013). Failure to consider psychosocial factors can lead to valuable objective information being missed.

For this reason, every figure involved both in the decision-making choice and in the request for an advisory opinion must necessarily have a deep knowledge of the social model within which they operate. In this regard, sharing information inside the stakeholders' group is of fundamental importance. Furthermore, DMP optimization must include a continuous exchange of information and a possible reformulation/revision of the rehabilitation project in the stakeholders' group through brainstorming (Ardern et al. 2016c). Psychological factors include apprehension, fear, and anxiety. In addition to negatively interfering with performance, these parameters represent a risk factor for reinjury (McCarty et al. 2004; Bauman 2005; Clover and Wall 2010; Delvaux et al. 2014). Therefore, we pointed out the following:

1. During RTT and RTP, decisions should take into account the psychological state of the athlete (McCarty et al. 2004; Bauman 2005; Diehl et al. 2006; Glazer 2009; Langford et al. 2009).
2. Individuals such as the coach, technical staff, and others may exert pressure on the RTT and RTP decision-making process (Tucker 2004; Best and Brolinson 2005; Bauman 2005; Burnett and Sonntag 2006; Verrall et al. 2006; Tator 2008; Langford et al. 2009¸ Creighton et al. 2010; Matheson et al. 2011). Despite these figures may bring important additional information to the DMP, sometimes they can disturb it heavily. Indeed, a potential conflict of interest exists between the athlete's needs and the wishes of the coach, the technical staff, and/or the management team of the club (Maron et al. 1994; Tucker 2004¸ Fuller and Walker 2006). We recommend that all stakeholders are safeguarded from external pressure in order to maintain maximum objectivity during RTT and RTP decisions.
3. The decision-making process must be based on a continual exchange of information between all stakeholders. This should allow for reformulation/revision of the rehabilitation plan when necessary (Ardern et al. 2016a, b, c).
4. The RTT and RTP decision-making process must be based on a continuum that runs parallel to the rehabilitation process. Isolated decisions regarding RTT and RTP that are not part of the rehabilitation process are to be avoided (Ardern et al. 2016a, b, c).
5. The RTT and RTP decision-making process must be "player-centered." The central role of the player/patient is to be respected by taking the following into account:
   (a) The short-term, medium-term, and long-term health risks associated with RTT and RTP
   (b) The role of player/patient as an active "decision maker" when deciding whether to RTT or RTP

2.16 Conclusions

Despite being a *continuum* in the decision-making process for the return to sporting activity, RTT and RTP are based on different judgment criteria: the former is based on clinical-functional criteria, while the latter is essentially based on performance criteria. Furthermore, some tests for the RTT presented in the following chapters can be considered as general, in the sense that they can also adapt to different muscle injuries, while others are highly muscle injury type-specific. Such specificity is a basic concept in the RTT-DMP. On the contrary, the tests on which the RTP is based are general because they check the performance once the RTT step has been freed successfully.

References

Almeida AM, Santos Silva PR, Pedrinelli A, Hernandez AJ. Aerobic fitness in professional soccer players after anterior cruciate ligament reconstruction. PLoS One. 2018;13(3):e0194432.

Ardern CL, Glasgow P, Schneiders A, et al. 2016 consensus statement on return to sport from the first world congress in sports physical therapy, Bern. Br J Sports Med. 2016a;50:853–64.

Ardern CL, Bizzini M, Bahr R. It is time for consensus on return to play after injury: five key questions. Br J Sports Med. 2016b;50:506–8.

Ardern CL, Glasgow P, Schneiders A, Witvrouw E, Clarsen B, Cools A, Gojanovic B, Griffin S, Khan KM, Moksnes H, Mutch SA, Phillips N, Reurink G, Sadler R, Silbernagel KG, Thorborg K, Wangensteen A, Wilk KE, Bizzini M. 2016 Consensus statement on return to sport from the First World Congress in Sports Physical Therapy, Bern. Br J Sports Med. 2016c;50(14):853–64.
Atkins E, Colville G, John M. A 'biopsychosocial' model for recovery: a grounded theory study of families' journeys after a Paediatric Intensive Care admission. Intensive Crit Care Nurs. 2012;28:133–40.
Ayers DC, Franklin PD, Ring DC. The role of emotional health in functional outcomes after orthopaedic surgery: extending the biopsychosocial model to orthopaedics. J Bone Joint Surg Am. 2013;95:e165.
Badawy MM, Muaidi QI. Aerobic profile during high-intensity performance in professional Saudi athletes. Pak J Biol Sci. 2018;21(1):24–8.
Bauman J. Returning to play: the mind does matter. Clin J Sport Med. 2005;15:432–5.
Bell NS, Mangione TW, Hemenway D, et al. High injury rates among female army trainees: a function of gender? Am J Prev Med. 2000;18(suppl 3):141–6.
Best TM, Brolinson PG. Return to play: the sideline dilemma. Clin J Sport Med. 2005;15:403–4.
Bisciotti GN, Volpi P. Return to play. In: Volpi P, editor. Football doctor manual. Trento: Edra Edition; 2018. p. 247–59.
Bisciotti GN, Volpi P, Amato M, Alberti G, Allegra F, Aprato A, Artina M, Auci A, Bait C, Bastieri GM, Balzarini L, Belli A, Bellini G, Bettinsoli P, Bisciotti A, Bisciotti A, Bona S, Brambilla L, Bresciani M, Buffoli M, Calanna F, Canata GL, Cardinali D, Carimati G, Cassaghi G, Cautero E, Cena E, Corradini B, Corsini A, D'Agostino C, De Donato M, Delle Rose G, Di Marzo F, Di Pietto F, Enrica D, Eirale C, Febbrari L, Ferrua P, Foglia A, Galbiati A, Gheza A, Giammattei C, Masia F, Melegati G, Moretti B, Moretti L, Niccolai R, Orgiani A, Orizio C, Pantalone A, Parra F, Patroni P, Pereira Ruiz MT, Perri M, Petrillo S, Pulici L, Quaglia A, Ricciotti L, Rosa F, Sasso N, Sprenger C, Tarantola C, Tenconi FG, Tosi F, Trainini M, Tucciarone A, Yekdah A, Vuckovic Z, Zini R, Chamari K. Italian consensus conference on guidelines for conservative treatment on lower limb muscle injuries in athlete. BMJ Open Sport Exerc Med. 2018;4(1):e000323.
Bowen TR, Feldmann DD, Miller MD. Return to play following surgical treatment of meniscal and chondral injuries to the knee. Clin Sports Med. 2004;23:381–93.
Bringard A, Pogliaghi S, Adami A, De Roia G, Lador F, Lucini D, Pizzinelli P, Capelli C, Ferretti G. Cardiovascular determinants of maximal oxygen consumption in upright and supine posture at the end of prolonged bed rest in humans. Respir Physiol Neurobiol. 2010;172(1–2):53–62.
Burnett MG, Sonntag VKH. Return to contact sports after spinal surgery. Neurosurg Focus. 2006;21:1–3.
Cantu RC. Cervical spine injuries in the athlete. Semin Neurol. 2000;20:173–8.
Chamari K, Hachana Y, Ahmed YB, Galy O, Sghaïer F, Chatard JC, Hue O, Wisløff U. Field and laboratory testing in young elite soccer players. Br J Sports Med. 2004;38(2):191–6.
Chamari K, Moussa-Charai I, Boussaïdi L, Hachana Y, Kauech F, Wisløff U. Appropriate interpretation of aerobic capacity: allometric scaling in adult and young soccer players. Br J Sports Med. 2005;39(2):97–101.
Chomiak J, Junge A, Peterson L, et al. Severe injuries in football players. Influencing factors. Am J Sports Med. 2000;28(Suppl 5):S58–68.
Clover J, Wall J. Return-to-play criteria following sports injury. Clin Sports Med. 2010;29:169–75.
Convertino VA. Cardiovascular consequences of bed rest: effect on maximal oxygen uptake. Med Sci Sports Exerc. 1997;29(2):191–6.
Creighton DW, Shrier I, Shultz R, et al. Return-to-play in sport: a decision-based model. Clin J Sport Med. 2010;20:379–85.
Curl LA. Return to sport following elbow surgery. Clin Sports Med. 2004;23:353–66.
Delvaux F, Rochcongar P, Bruyere O, et al. Return-to-play criteria after hamstring injury: actual medicine practice in professional soccer teams. J Sports Sci Med. 2014;13:721–3.
Diehl JJ, Best TM, Kaeding CC. Classification and return-to-play considerations for stress fractures. Clin Sports Med. 2006;25:17–28.
Dimberg EL, Burns TM. Management of common neurologic conditions in sports. Clin Sports Med. 2005;24:637–62.
Drake DF, Nadler SF, Chou LH, et al. Sports and performing arts medicine. 4. Traumatic injuries in sports. Arch Phys Med Rehabil. 2004;85:S67–71.
Dunn IF, Proctor MR, Day AL. Lumbar spine injuries in athletes. Neurosurg Focus. 2006;21:E4.
Dvorak J, Junge A, Chomiak J, et al. Risk factor analysis for injuries in football players. Possibilities for a prevention program. Am J Sports Med. 2000;28(Suppl 5):S69–74.
Eck JC, Riley LH III. Return to play after lumbar spine conditions and surgeries. Clin Sports Med. 2004;23:367–79.
Eddy D, Congeni J, Loud K. A review of spine injuries and return to play. Clin J Sport Med. 2005;15:453–8.
Elias I, Pahl MA, Zoga AC, et al. Recurrent burner syndrome due to presumed cervical spine osteoblastoma in a collision sport athlete—a case report. J Brachial Plex Peripher Nerve Inj. 2007;2:13.
Fuller CW, Walker J. Quantifying the functional rehabilitation of injured football players. Br J Sports Med. 2006;40:151–7.
Gabbett TJ, Benton D. Reactive agility of rugby league players. J Sci Med Sport. 2009;12:212–4.
Glazer DD. Development and preliminary validation of the injury- psychological readiness to return to sport (I-PRRS) scale. J Athl Train. 2009;44:185–9.
Heck H, Mader A, Hess G, Mucke S, Muller R, Hollmann W. Justification of the 4-mmol/l lactate threshold. Int J Sports Med. 1985;6:117–30.

Herring SA, Bergfeld JA, Boyd J, Duffey T, Fields KB, Grana WA, Indelicato P, Kibler WB, Pallay R, Putukian M, Sallis RE. The team physician and return-to-play issues: a consensus statement. Med Sci Sports Exerc. 2002;34(7):1212–4.

Herring SA, Kibler WB, Putukian M. The team physician and the return-to-play decision: a consensus statement-2012 update. Med Sci Sports Exerc. 2012;44(12):2446–8.

Herring SA, Kibler WB, Putukian M. The team physician consensus statement: 2013 update. Med Sci Sports Exerc. 2013;45(8):1618–22.

Hopper DM, Hopper JL, Elliott BC. Do selected kinanthropometric and performance variables predict injuries in female netball players? J Sports Sci. 1995;13:213–22.

Kaeding CC, Yu JR, Wright R, et al. Management and return to play of stress fractures. Clin J Sport Med. 2005;15:442–7.

Kovacic J, Bergfeld J. Return to play issues in upper extremity injuries. Clin J Sport Med. 2005;15: 448–52.

Krabak B, Kennedy DJ. Functional rehabilitation of lumbar spine injuries in the athlete. Sports Med Arthrosc. 2008;16:47–54.

Kuhn JE. Treating the initial anterior shoulder dislocation—an evidencebased medicine approach. Sports Med Arthrosc. 2006;14:192–8.

Langford JL, Webster KE, Feller JA. A prospective longitudinal study to assess psychological changes following anterior cruciate ligament reconstruction surgery. Br J Sports Med. 2009;43:377–8.

Lockie RG, Jeffriess MD, McGann TS, et al. Planned and reactive agility performance in semiprofessional and amateur basketball players. Int J Sports Physiol Perform. 2014;9:766–71.

Lord J, Winell JJ. Overuse injuries in pediatric athletes. Curr Opin Pediatr. 2004;16:47–50.

Maron BJ, Brown RW, McGrew CA, et al. Ethical, legal, and practical considerations impacting medical decision-making in competitive athletes. Med Sci Sports Exerc. 1994;26(Suppl):S230–7.

Matheson GO, Shultz R, Bido J, et al. Return-to-play decisions: are they the team physician's responsibility? Clin J Sport Med. 2011;21:25–30.

McCarty EC, Ritchie P, Gill HS, et al. Shoulder instability: return to play. Clin Sports Med. 2004;23:335–51.

Miller MD, Arciero RA, Cooper DE, et al. Doc, when can he go back in the game? Instr Course Lect. 2009;58:437–43.

Murphy DF, Connolly DA, Beynnon BD. Risk factors for lower extremity injury: a review of the literature. Br J Sports Med. 2003;37(1):13–29.

Orchard J, Best TM, Verrall GM. Return to play following muscle strains. Clin J Sport Med. 2005;15:436–41.

Ostenberg A, Roos H. Injury risk factors in female European football. A prospective study of 123 players during one season. Scand J Med Sci Sports. 2000;10:279–85.

Park HB, Lin SK, Yokota A, et al. Return to play for rotator cuff injuries and superior labrum anterior posterior (SLAP) lesions. Clin Sports Med. 2004;23:321–34.

Pincus T, Kent P, Bronfort G, et al. Twenty-five years with the biopsychosocial model of low Back pain—is it time to celebrate? A report from the Twelfth International Forum for Primary Care Research on Low Back Pain. Spine. 2013;38:2118–23.

Saal JA. Common American football injuries. Sports Med. 1991;12:132–47.

Sheppard JM, Young WB, Doyle TLA, et al. An evaluation of a new test of reactive agility and its relationship to sprint speed and change of direction speed. J Sci Med Sport. 2006;9:342–9.

Shrier I. Strategic Assessment of Risk and Risk Tolerance (StARRT) framework for return-to-play decision-making. Br J Sports Med. 2015;49:1311–5.

Smurawa T, Congeni J. Return-to-play decisions in the adolescent athlete: how to decide. Pediatr Ann. 2007;36:746–8. 750–741.

Tator CH. Recognition and management of spinal cord injuries in sports and recreation. Neurol Clin. 2008;26:79–88.

Tucker AM. Ethics and the professional team physician. Clin Sports Med. 2004;23:227–41.

Verrall GM, Brukner PD, Seward HG. 6. Doctor on the sidelines. Med J Aust. 2006;184:244–8.

Weiler R. Unknown unknowns and lessons from non-operative rehabilitation and return to play of a complete anterior cruciate ligament injury in English Premier League football. Br J Sports Med. 2016;50:261–2.

Wiese-Bjornstal DM, Smith AM, Shaffer SM, et al. An integrated model of response to sport injury: psychological and sociological dynamics. J Appl Sport Psychol. 1998;10:46–69.

3 The Structure of the Return to Training Decision-Making Process

3.1 Introduction to RTT Assessment

The RTT decision-making process (DMP) is grounded on four different types of assessment:

1. Clinical assessment
2. Imaging
3. Laboratory test
4. Field tests

The above hierarchy in the structure of the RTT-DMP allows to safely test the player by minimizing the risk of reinjury. Indeed, especially field tests represent the moment a player risks reinjury the most. For this reason, maximum prudence must always guide the structuration of the tests.

3.2 Clinical Assessment

The clinical assessment is structured through general tests (i.e., suitable for every muscle group) and specific tests that are designed for each considered muscle group. The specific tests will be discussed in the chapters focused on different considered muscle groups, while the suggested general tests are:

1. Absence of clinical symptoms (Kvist 2004; Malliaropoulos et al. 2011; Delvaux et al. 2014)
2. Absence of pain or tenderness during muscle palpation (Kvist 2004; Delvaux et al. 2014; Reurink et al. 2014; Zambaldi et al. 2017)
3. Absence of pain on passive and active stretching (Witvrouw et al. 2003; Bisciotti et al. 2018)
4. Absence of pain on isometric, concentric, and eccentric contraction (Bisciotti et al. 2018)
5. Completion of the prescribed rehabilitation program (Reurink et al. 2014)
6. MRI and US imaging assessment in compliance with the points specified in paragraph "The role of imaging in the RTT-DMP" (Connell et al. 2004; Sanfilippo et al. 2013; Reurink et al. 2014)

Furthermore, please note that during clinical assessment, the subjective feelings of the player are taken into account (i.e., assessing levels of anxiety, apprehension, fear of failure, and/or fear of reinjury) (McCarty et al. 2004; Bauman 2005; Clover and Wall 2010; Delvaux et al. 2014).

3.3 The Role of Imaging in the RTT-DMP

Many studies report how, following indirect muscle injury, at the time of RTP (ranging from 29 to 49 days), 50% to 90% of subjects still showed

G. N. Bisciotti et al., *Return-to-Play after Lower Limb Muscle Injury in Football*,
https://doi.org/10.1007/978-3-030-84950-4_3

signs of MRI signal alteration (i.e., hyperintensity in sensitive fluid sequences) (Connell et al. 2004; Sanfilippo et al. 2013; Reurink et al. 2014). Furthermore, 32% of the subjects in the same period show some alteration at the US examination (Connell et al. 2004). In these studies, the area under the anomalous MRI signal, at the time of RTP, ranged on average from 20% to 28% of the area measured at the baseline, that is, at the time of the injury. On the other hand, there are no reliable data regarding the area of alteration of the US signal at RTP (Connell et al. 2004). Both the MRI and the US signals normalized after an average of 6 months (Connell et al. 2004; Sanfilippo et al. 2013; Reurink et al. 2014). Despite persistent alteration, the percentage of reinjuries was less than 2% (Connell et al. 2004; Sanfilippo et al. 2013; Reurink et al. 2014). The presence of abnormalities on MRI and US during this period may find an explanation in the greater number of the ionic interactions of immature collagen formed during the early stage of muscle healing. The conversion of these weaker bonds to stronger covalent bonds, during posttranslational modifications of the constituent amino acids, may require longer periods of up to 6 months depending on the extent of the injury (Connell et al. 2004). A further important point to underline is that several studies of post-lesion tissue at the time of RTP demonstrate that 34% of athletes exhibit a low-intensity MRI signal, indicative of the formation of fibrotic scar tissue (Connell et al. 2004; Bedair et al. 2008; Gharaibeh et al. 2012). Since the formation of an excessive area of fibrotic tissue leads to an alteration of muscular stiffness, this situation must be interpreted as a risk factor for reinjury (Connell et al. 2004; Bedair et al. 2008; Gharaibeh et al. 2012).

Therefore, in respect of imaging, the Consensus Conference specifies the following:

1. The RTT decision-making process does not necessarily require a total resolution of MRI and US area of signal alteration (Connell et al. 2004; Sanfilippo et al. 2013; Reurink et al. 2014).
2. In MRI, a signal alteration (hyperintensity zone in fluid-sensitive sequences) decreased by at least 70% in comparison with the baseline signal alteration is acceptable for RTT. This is justified by the fact that a decrease of at least 70% of the area, or of the volume, of hyperintensity signal in comparison to the baseline is not associated to reinjuries (Slavotinek 2010).
3. The presence of an extensive area of low signal intensity, indicative of fibrotic scar tissue, must be interpreted as a risk factor for reinjury (Connell et al. 2004; Bedair et al. 2008; Gharaibeh et al. 2012). However, attention must be paid to the fact that a hemosiderin deposition, following hemorrhage, can mimic the formation of fibrotic tissue (Slavotinek 2010).
4. The CSA derived from a fat-suppressed T2 in axial projection should be calculated as follows:

$$0.25\pi\,\mathrm{ML} * \mathrm{AP}$$

where ML is the medial-lateral distance and AP the anteroposterior distance of the area of hyperintensity. It should be noted that it is also possible to calculate the volume of the entire hyperintensity zone by adding the different CSAs and multiplying this value by the inter-slice distance. The measure must be performed at level of the maximum extension of the signal hyperintensity.

5. Given its greater sensitivity and greater tissue contrast gradient, MRI is preferable to US when making RTT decisions (Slavotinek 2010; Connell et al. 2004).

3.4 Laboratory Tests

The laboratory tests are based on the strength evaluation performed in isometric, isotonic, and isokinetic technology (Croisier et al. 2002; Delvaux et al. 2014; Sanfilippo et al. 2013; Bisciotti et al. 2016).

The isometric test should respect the following points (Webber and Porter 2010; Toonstra and Mattacola 2013; Ruschel et al. 2015):

1. Operate a proper warm-up.
2. Biomechanically isolate the muscle group to be tested.

3. Standardize the lever arm.
4. Begin the test with the healthy limb.
5. Apply an isometric contraction of progressive intensity for a duration of between 3 s and 5 s.
6. Encourage the patient during the test.
7. Perform at least three trials with an adequate recovery between each trial (around 1 min 30 s).
8. Consider the peak value.
9. Check for any pain symptoms with VAS.
10. Stop the test in the presence of severe pain (VAS >3).
11. The dynamometric values must be ≥90% of the pre-injury values an ≥90% of the contralateral limb values.

The isotonic test should respect the following points (Webber and Porter 2010; Van Driessche et al. 2018):

1. Operate a proper warm-up.
2. Biomechanically isolate the muscle group to be tested.
3. Standardize the lever arm (if possible) and ROM.
4. Begin the test with the healthy limb.
5. Apply the maximum speed during the movement.
6. Encourage the patient during the test.
7. Perform at least one set of six to ten repetitions;
8. Consider both average and peak value of force and power production.
9. Check for any pain symptoms with VAS.
10. Stop the test in the presence of severe pain (VAS >3).
11. The dynamometric values must be ≥90% of the pre-injury values and ≥ 90% of the contralateral limb values.

The isokinetic test should respect the following points (Croisier et al. 2002; Orchard et al. 2005; Delvaux et al. 2014; Sanfilippo et al. 2013; Bisciotti et al. 2016):

1. Operate a proper warm-up.
2. Biomechanically isolate the muscle group to be tested.
3. Standardize the lever arm and ROM.
4. Align the center of rotation of the joint with that of the mechanical device.
5. Subtract the weight of the limb from the calculation of the force moment (usually done automatically by the device).
6. Consider the average value avoiding the so-called "peak artifact."
7. Check for any pain symptoms with VAS.
8. Stop the test in the presence of severe pain (VAS >3).
9. Perform one set of six to ten repetitions at slow speed (30°–60°/s) and one set at high speed (≥300°/s).
10. Perform at least one eccentric test at 60°/s or 30°/s.
11. Begin the test with the healthy limb.
12. Encourage the patient during the test.
13. Perform the test observing an adequate recovery between the sets (~2–3′);
14. Consider the value of the joint angle corresponding to the peak of force production (during both concentric and eccentric tests). In the injured limb, the angle value in which the peak of force production is recorded should not differ more than 10% from the healthy limb.
15. Consider the value of mechanical work.
16. Consider the shape of the force curve.
17. Consider the value of the ratio of hamstring (concentric modality) to quadriceps (concentric modality) and the value of the ratio of hamstring (eccentric modality) to quadriceps (concentric modality). Concerning the hamstring injury, an altered ratio hamstring concentric/quadriceps concentric/Q (i.e., <0.65) and an altered ratio hamstring eccentric/quadriceps concentric (i.e., <1) should be considered as reinjury risk factors.
18. The dynamometric values must be ≥90% of the pre-injury values and ≥90% of the contralateral limb values.

References

Bauman J. Returning to play: the mind does matter. Clin J Sport Med. 2005;15:432–5.

Bedair HS, Karthikeyan T, Quintero A, et al. Angiotensin II receptor blockade administered after injury improves muscle regeneration and decreases fibrosis in normal skeletal muscle. Am J Sports Med. 2008;36:1548–54.

Bisciotti GN, Quaglia A, Belli A, Carimati G, Volpi P. Return to sports after ACL reconstruction: a new functional test protocol. Muscles Ligaments Tendons J. 2016;6(4):499–509.

Bisciotti GN, Volpi P, Amato M, Alberti G, Allegra F, Aprato A, Artina M, Auci A, Bait C, Bastieri GM, Balzarini L, Belli A, Bellini G, Bettinsoli P, Bisciotti A, Bisciotti A, Bona S, Brambilla L, Bresciani M, Buffoli M, Calanna F, Canata GL, Cardinali D, Carimati G, Cassaghi G, Cautero E, Cena E, Corradini B, Corsini A, D'Agostino C, De Donato M, Delle Rose G, Di Marzo F, Di Pietto F, Enrica D, Eirale C, Febbrari L, Ferrua P, Foglia A, Galbiati A, Gheza A, Giammattei C, Masia F, Melegati G, Moretti B, Moretti L, Niccolai R, Orgiani A, Orizio C, Pantalone A, Parra F, Patroni P, Pereira Ruiz MT, Perri M, Petrillo S, Pulici L, Quaglia A, Ricciotti L, Rosa F, Sasso N, Sprenger C, Tarantola C, Tenconi FG, Tosi F, Trainini M, Tucciarone A, Yekdah A, Vuckovic Z, Zini R, Chamari K. Italian consensus conference on guidelines for conservative treatment on lower limb muscle injuries in athlete. BMJ Open Sport Exerc Med. 2018;4(1):e000323.

Clover J, Wall J. Return-to-play criteria following sports injury. Clin Sports Med. 2010;29:169–75.

Connell DA, Schneider-Kolsky ME, Hoving JL, Malara F, Buchbinder R, Koulouris G, Burke F, Bass C. Longitudinal study comparing sonographic and MRI assessments of acute and healing hamstring injuries. AJR Am J Roentgenol. 2004;183(4):975–84.

Croisier J, Forthomme B, Namurois M, Vanderthommen M, Crielaard J. Hamstring muscle strain recurrence and strength performance disorders. Am J Sports Med. 2002;30(2):199–203.

Delvaux F, Rochcongar P, Bruyere O, et al. Return-to-play criteria after hamstring injury: actual medicine practice in professional soccer teams. J Sports Sci Med. 2014;13:721–3.

Gharaibeh B, Chun-Lansinger Y, Hagen T, et al. Biological approaches to improve skeletal muscle healing after injury and disease. Birth Defects Res C Embryo Today. 2012;96:82–94.

Kvist J. Rehabilitation following anterior cruciate ligament injury: current recommendations for sport participation. Sports Med. 2004;34:296–80.

Malliaropoulos N, Isinkaye T, Tsitas K, et al. Reinjury after acute posterior thigh muscle injuries in elite track and field athletes. Am J Sports Med. 2011;39(2):304–10.

McCarty EC, Ritchie P, Gill HS, et al. Shoulder instability: return to play. Clin Sports Med. 2004;23:335–51.

Orchard J, Best TM, Verrall GM. Return to play following muscle strains. Clin J Sport Med. 2005;15:436–41.

Reurink G, Goudswaard GJ, Tol JL, et al. MRI observations at return to play of clinically recovered hamstring injuries. Br J Sports Med. 2014;48:1370–6.

Ruschel C, Haupenthal A, Jacomel GF, Fernandes Jacomel G, et al. Validity and reliability of an instrumented leg-extension machine for measuring isometric muscle strength of the knee extensors. J Sport Rehabil. 2015;24:2013–122.

Sanfilippo JL, Silder A, Sherry MA, Tuite MJ, Heiderscheit BC. Hamstring strength and morphology progression after return to sport from injury. Med Sci Sports Exerc. 2013;45(3):448–54.

Slavotinek JP. Muscle injury: the role of imaging in prognostic assignment and monitoring of muscle repair. Semin Musculoskelet Radiol. 2010;14:194–200.

Toonstra J, Mattacola CG. Test-retest reliability and validity of isometric knee-flexion and -extension measurement using 3 methods of assessing muscle strength. J Sport Rehabil. 2013;22(1) https://doi.org/10.1123/jsr.2013.TR7.

Van Driessche S, Delecluse C, Bautmans I, et al. Age-related differences in rate of power development exceed differences in peak power. Exp Gerontol. 2018;101:95–100.

Webber SC, Porter MM. Reliability of ankle isometric, isotonic, and isokinetic strength and power testing in older women. Phys Ther. 2010;90:1165–75.

Witvrouw E, Danneels L, Asselman P, et al. Muscle flexibility as a risk factor for developing muscle injuries in male professional soccer players. Am J Sports Med. 2003;31:41–6.

Zambaldi M, Beasley I, Rushton A. Return to play criteria after hamstring muscle injury in professional football: a Delphi consensus study. Br J Sports Med. 2017;51:1221–6.

4 The Structure of the Return to Play Decision-Making Process

4.1 The RTP Basic Principles

As already specified, the evaluation of RTP is a judgment of suitability for the resumption of full training load without restrictions, as well as of the ability for the player to conveniently face the competition.

The RTP-DMP makes use of the GPS technology which allows to collect and analyze the indices on which the RTP-DMP is based (Colby et al. 2014; Gleason et al. 2017; Barr et al. 2019).

The foundational principles of RTP-DMP can be divided into three evaluation categories:

1. Quantitative evaluation
2. Qualitative evaluation
3. Parameter analysis

4.2 Quantitative Evaluation (QNE)

The QNE (Di Prampero et al. 2005, 2015; Osgnach et al. 2010; Bisciotti and Volpi 2018) requires the analysis of speed (divided into six progressively increasing speed categories) recorded in the last period of pre-injury training versus the same parameters recorded in the post-injury period between RTT and the formulation of the judgment of suitability for the RTP.

For the judgment to be based on superimposable indices, at least in general, the sampling should be carried out in training sessions that are as similar as possible to each other. The calculation must proceed through the numerical and graphic representation (in order to be immediately understandable) of each of the six categories of speed listed, and the recorded data should account for the time spent and distance covered at the indicated velocity. The suggested speed categories are:

1. Walking (range 0 to <5.4 km/h)
2. Jogging (range 5.5 to <10.8 km/h)
3. Low speed running (range 10.9 to <14.4 km/h)
4. Intermediate speed running (range 14.5 to <19.8 km/h)
5. High speed running (range 19.9 to <25.2 km/h)
6. Maximum speed running (≥ 25.2 km/h)

An example and graphic representation of QNE is shown in Fig. 4.1.

4.3 Qualitative Evaluation (QLE)

QLE is based on the analysis of metabolic power (MP) calculated with GPS technology. The MP (expressed in W/kg^{-1}) represents the product of speed and acceleration in determining the inten-

G. N. Bisciotti et al., *Return-to-Play after Lower Limb Muscle Injury in Football*,
https://doi.org/10.1007/978-3-030-84950-4_4

Speed ranges	Pre-injury period (%)	After RTT period (%)
Walking	100	95
Jogging	100	92
Low speed running	100	87
Intermediate speed running	100	84
Hight speed running	100	72
Maximum speed running	100	67

Speed ranges	Pre-injury period (%)	After RTT period (%)
Walking	100	100
Jogging	100	95
Low speed running	100	90
Intermediate speed running	100	87
Hight speed running	100	78
Maximum speed running	100	73

Speed ranges	Pre-injury period (%)	After RTT period (%)
Walking	100	100
Jogging	100	100
Low speed running	100	100
Intermediate speed running	100	95
Hight speed running	100	98
Maximum speed running	100	93

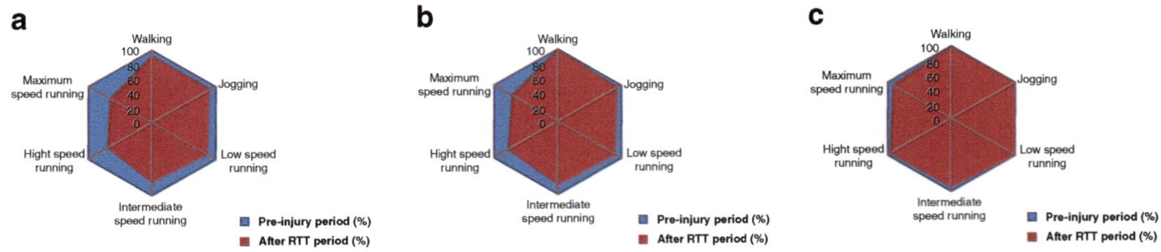

Fig. 4.1 The QNE is based on the analysis of the velocity ranges recorded during the pre-injury period compared to the same parameters recorded during the period following RTT and previous to RTP. The parameters recorded during the pre-injury period are all arbitrarily set to 100. In (**a–c**), the graphic representation shows the progressive overlapping of the areas related to the two data acquisition periods: the greater the overlap of the parameters, the closer the return to pre-injury level of performance (Bisciotti and Volpi 2018)

sity of running (Di Prampero et al. 2005, 2015; Osgnach et al. 2010).

The MP value can be calculated using the following formula:

$$MP = CE \cdot v$$

In which CE represents the energy cost of running at a constant speed (equal to 1 kcal·kg^{-1}·km^{-1}) and *v* is the athlete's instantaneous speed. Below is the suggested division of MP into six progressively greater categories. MP is calculated by quantifying the time spent in each MP category.

1. Low power (0 to <5 W. kg^{-1})
2. Intermediate power (5.1 to <10 W. kg^{-1})
3. High power (10.1 to <20 W. kg^{-1})
4. Higher power (20.1 to <25 W. kg^{-1})
5. Very high power (25.1 to <50 W. kg^{-1})
6. Maximum power (≥50 W. kg^{-1})

MP time values recorded in the last period of pre-injury training are compared, as in the case of QNE, to the same parameters recorded post-injury to formulate the RTP judgment.

An example and graphic representation of QLE is shown in Fig. 4.2.

4.4 Parameter Analysis (PA)

The PA consists in the comparison of a certain number of parameters recorded during the pre-injury period versus the same parameters collected during the period between RTT and RTP. The PA is based on the following data (Di Prampero et al. 2005, 2015; Osgnach et al. 2010):

1. Total distance covered during training (regardless of the run speed).
2. The equivalent distance (ED). In football, energy expenditure is influenced by the accelerating and decelerating components of the activity (Osgnach et al. 2010; Di Prampero et al. 2015). The ED corresponds to the distance that the athlete could theoretically cover if he ran, at constant speed, using the same total energy expenditure as that used during the game. The ED value can be calculated using the following formula (Di Prampero et al. 2005):

$$ED = W / Ecc$$

where *W* represents the energy cost expressed in Jkg^{-1} and Ecc is the energy cost of running

Metabolic power ranges	Pre-injury period(%)	After RTT period (%)
Low power	100	94
Intermediate power	100	91
High power	100	86
Higher power	100	83
Very high power	100	71
Maximum power	100	66

Metabolic power ranges	Pre-injury period(%)	After RTT period (%)
Low power	100	99
Intermediate power	100	94
High power	100	89
Higher power	100	86
Very high power	100	77
Maximum power	100	72

Metabolic power ranges	Pre-injury period(%)	After RTT period (%)
Low power	100	100
Intermediate power	100	100
High power	100	99
Higher power	100	94
Very high power	100	97
Maximum power	100	93

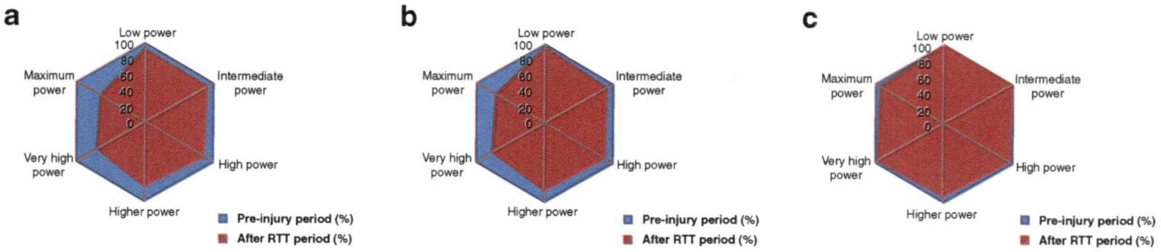

Fig. 4.2 QLE is based on the analysis of the MP recorded during the pre-injury period as compared to the same parameters recorded during the period following RTT and previous to RTP. As seen with QNE, the parameters recorded during the pre-injury period are all arbitrarily set to 100. In (**a–c**), the graphic representation shows the progressive overlapping of the areas related to the two data acquisition periods. Likewise, a progressive growth of the overlapping of the QLE parameters indicates that the player is progressively returning to the pre-injury level of performance (Bisciotti and Volpi 2018)

in a straight line at constant speed on compact grassland (i.e., 4.6 J kg^{-1}).

3. The equivalent distance index (EDI). The EDI represents the ratio between the value of ED and the distance actually covered by the player (RD) according to the following formula (Di Prampero et al. 2005):

$$\mathrm{EDI} = \mathrm{ED} / \mathrm{RD}$$

4. The anaerobic index (AI). The AI represents the ratio between the energy cost beyond a certain metabolic threshold (i.e., anaerobic threshold value, or maximal aerobic speed value) and the total energy expenditure and is calculated as follows (Di Prampero et al. 2005):

$$\mathrm{AI} = \mathrm{Wtp} / W$$

where Wtp represents the energy consumed beyond the metabolic threshold considered (anaerobic threshold or maximal aerobic speed) expressed in $J{\cdot}kg^{-1}$ and W is the total energy expenditure, also expressed in $J{\cdot}kg^{-1}$.

References

Barr M, Beaver T, Turczyn D, Cornish S. Validity and reliability of 15 Hz global positioning system units for assessing the activity profiles of university football players. J Strength Cond Res. 2019;33(5):1371–9.

Bisciotti GN, Volpi P. Return to play. In: Volpi P, editor. Football doctor manual. Trento: Edra Edition; 2018. p. 247–59.

Colby MJ, Dawson B, Heasman J, Rogalski B, Gabbett TJ. Accelerometer and GPS-derived running loads and injury risk in elite Australian footballers. J Strength Cond Res. 2014;28(8):2244–52.

Di Prampero PE, Fusi S, Sepulcri L, Morin JB, Belli A, Antonutto G. Sprint running a new energetic approach. J Exp Biol. 2005;208(Pt 14):2809–16.

Di Prampero PE, Botter A, Osgnach C. The energy cost of sprint running and the role of metabolic power in setting top performances. Eur J Appl Physiol. 2015;115(3):451–69.

Gleason BH, Sams ML, Salley JT, Pustina AA, Stone MH. Global positioning system analysis of a high school football scrimmage. J Strength Cond Res. 2017;31(8):2183–8.

Osgnach C, Poser S, Bernardini R, Rinaldo R, di Prampero PE. Energy cost and metabolic power in elite soccer: a new match analysis approach. Med Sci Sports Exerc. 2010;42(1):170–8.

5 The Role of Imaging in the Return to Training and Return to Play Decision-Making Process

5.1 Introduction

Magnetic resonance imaging (MRI) is a validated tool for diagnosis and prognosis of muscle injuries and is widely used in sport injuries, especially in the elite athletes (Connell et al. 2004; Comin et al. 2013; Kerkhoffs et al. 2013). Even if the follow-up MRI examination has been recommended by several authors to monitor the recovery after muscle injury and to support the subsequent return to training decision-making process, the MRI examination procedure and its data interpretation has not been validated yet (Mendiguchia and Brughelli 2011; Kerkhoffs et al. 2013). Indeed, despite a correct approach based on conventional clinical and imaging examination, the return to play decision-making process remains a challenge (Askling et al. 2010; Heiderscheit et al. 2010). This is confirmed by the high number of muscle reinjures occurring after the return to training and/or return to play. Furthermore, 59% of muscle reinjures occurred in the first month after the return to play (Brooks et al. 2006). These data show the importance to have a reliable assessment tool able to discriminate between the athletes who are ready and those who are not, for return to training and return to play. It is clear that imaging examination plays a crucial role in the optimization of return to training decision-making process (Mendiguchia and Brughelli 2011; Kerkhoffs et al. 2013). However, in literature, there is no consensus concerning the role and the value of MRI examination in monitoring recovery and, consequently, in return to training and return to play decision (Reurink et al. 2014).

5.2 The Current State of Art

Many studies report that, within a range of 29–49 days after return to play following indirect muscle injury, from 50% to 90% of the subjects still showed signs of signal alteration on MRI examination, which consisted in hyperintensity in fluid-sensitive sequences (Connell et al. 2004; Sanfilippo et al. 2013; Reurink et al. 2014). Moreover, in the same period, 32% of the considered subjects showed alteration in the ultrasound (US) examination (Connell et al. 2004). On average, the area of alteration of the MRI signal at the time of return to play was between 20% and 28% of the area measured at the baseline, that is, at the time of the injury (Reurink et al. 2014). On the other hand, there is no reliable data regarding the area of alteration of the US signal at the time of the return to play (Connell et al. 2004). In the literature, there are some interesting studies on this topic.

Connell et al. (2004) considered 60 male professional football players with suspected acute hamstring strain. All the subjects underwent US and MRI examination within 3 days of injury; those who were injured underwent other US

G. N. Bisciotti et al., *Return-to-Play after Lower Limb Muscle Injury in Football*,
https://doi.org/10.1007/978-3-030-84950-4_5

and MRI examinations after 2 and 6 weeks. The results showed the presence of an increased signal intensity on MRI fluid-sensitive sequences, which was consistent with edema persisting after resolution of clinical symptoms, in the 36% of the considered subjects (15/42) who showed persistent abnormal findings on MRI examination 6 weeks after the injury. However, there were no reliable data regarding the area of alteration of the US signal at the time of the return to play. Such results led the authors to conclude that MRI is more sensitive than US for follow-up imaging of healing injuries. Furthermore, the longitudinal length of the injury measured on MRI is a strong predictor for the amount of time needed until an athlete can return to training and competition.

Askling et al. (2007) considered in case series 18 elite sprinters with acute first-time hamstring strains. All the subjects were examined, clinically and with magnetic resonance imaging, for four times after injury: at day 2 to 4, 10, 21, and 42. The clinical follow-up lasted 2 years. Six weeks after the injury, MRI examinations showed the presence of an increased signal intensity on fluid-sensitive sequences in 17 out of the 18 considered athletes. The authors concluded that the careful palpation of the injured area during the first 3 weeks after injury and MRI examinations performed during the first 6 weeks after injury provide valuable information that can be used to predict the time to return to pre-injury level of performance in elite sprinting athletes.

Silder et al. (2008) conducted a study in which MRI images were obtained from 14 athletes who had sustained a clinically diagnosed grade Ist–IInd hamstring strain injury between 5 and 23 months before, as well as from five healthy controls. The results showed that increased low-intensity signal was present along the musculotendon junction adjacent to the site of presumed prior injury for 11 of the 14 subjects. These data were suggestive of persistent scar tissue. Furthermore, two of these subjects also presented fatty infiltration within the previous injuries. In the authors' opinion, the result of the study showed evidence of long-term muscle-tendon remodeling following a muscle injury. Furthermore, since many athletes are likely returning to sport with residual hamstring atrophy, it is possible that long-term changes in muscle-tendon structure following injury alter contraction mechanics during functional request, which may increase the reinjury risk.

Sanfilippo et al. (2013) considered 25 athletes who experienced an acute hamstring strain injury following completion of a controlled rehabilitation program. Bilateral isokinetic strength testing and MRI examination were performed at return to play and 6 months later. The injured limb showed a peak torque deficit of 9.6% compared to the uninjured limb at return to play, but the same was not seen 6 months later. The MRI examination revealed that 20.4% of the muscle cross-sectional area showed signs of edema at return to play, and full resolution was only achieved after the 6-month follow-up. Furthermore, the tendon volume of the injured limb tended to increase over time ($p = 0.108$), while muscle volume decreased 4–5% in both limbs ($p < 0.001$). Basing on these results the authors concluded that residual edema and deficits in isokinetic knee flexion strength were present at the moment of return to sport activity but resolved during the subsequent 6 months. This happened despite the evidence of scar tissue formation (i.e., increased tendon volume) and muscle atrophy at MRI examination. Such data suggested that neuromuscular factors may contribute to the return to normal value of muscle strength.

Silder et al. (2008) in a randomized, double-blind, parallel group clinical trial considered 29 subjects who had suffered hamstring injury. The subjects were randomly assigned to two groups. The first group performed a rehabilitation program based on progressive agility and trunk stabilization, while the second group performed a rehabilitation program based on progressive running and eccentric strengthening. MRI and clinical examinations were conducted before and after completion of the rehabilitation programs. The recorded results showed few differences in clinical or morphological outcome measures between the two rehabilitation groups, and reinjury rates were low for both groups after return to sport (four reinjures out of 29 subjects). The cranio-caudal length of injury, measured on MRI examination before the start of rehabilitation, was positively correlated with the time for the return to sport. Furthermore, at the time of return to sport, no subject showed complete resolution of injury on MRI examination, even though all subjects were showing a near-complete resolution of pain and return of muscle strength. The authors concluded that evidence of continuing muscular healing is present after completion of rehabilitation, despite the appearance of normal physical strength and function on clinical examination.

In conclusion the literature shows that both RM signal and US signal normalized after a relatively long period of 6 months on average (Connell et al. 2004; Sanfilippo et al. 2013; Reurink et al. 2014). However, it is important to note that, despite the persistence of signal alteration, the percentage of reinjuries was less than 2% in all the considered studies. The presence of abnormalities on MRI and US during this period may be explained by the greater number of the ionic interactions of immature collagen formed during the early stage of muscle healing. The conversion of these weaker bonds to stronger covalent bonds, during posttranslational modifications of the constituent amino acids, may require longer periods of up to 6 months depending on the extent of the injury (Connell et al. 2004). A further aspect to consider is represented by some studies reporting that, at the time of return to play, 34% of athletes show, within the post-injured tissue reorganization zone, a low signal intensity, suggestive of fibrotic tissue formation. The presence of an extensive area of low signal intensity, indicative of fibrotic scar tissue, must be interpreted as a risk factor for reinjury (Bedair et al. 2008; Reurink et al. 2014, 2015a, b). However, attention must be paid to the fact that a hemosiderin deposition, following hemorrhage, can mimic the formation of fibrotic tissue (Slavotinek 2010). Since the formation of an excessive area of fibrotic tissue leads to an alteration of muscle stiffness, this situation must be interpreted as a risk factor for reinjury (Connell et al. 2004; Bedair et al. 2008; Gharaibeh et al. 2012).

5.3 The Italian Consensus Conference Suggestions

With respect to MRI and US imaging assessment, the following points must be specified:

1. The return to training decision-making process (and consequently the return to play decision-making process) does not necessarily require a total normalization of the RM and US signal (Connell et al. 2004; Sanfilippo et al. 2013; Reurink et al. 2014).
2. In the MRI examination, an alteration of the signal (i.e., the hyperintensity zone in fluid-sensitive sequences) should not be more than 30% compared to the baseline (Reurink et al. 2014; Sanfilippo et al. 2013). This is justified by the fact that a decrease of at least 70% of the area or volume of hyperintensity compared to the baseline, before the return to play, is not associated with reinjury (Slavotinek 2010). An example is visible in Figs. 5.1, 5.2, 5.3 and 5.4.
3. Despite some authors re-questioning this hypothesis (Reurink et al. 2015a, b), in the CC opinion, the presence of an extensive area of low signal intensity, suggestive for scarring

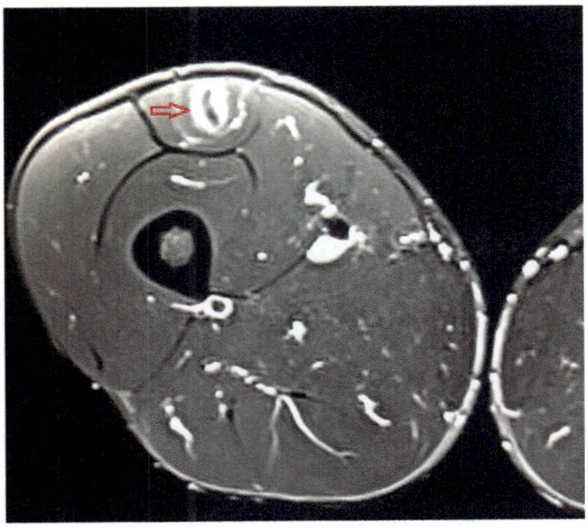

Fig. 5.1 MRI axial STIR of the middle third of the right thigh, showing, at the level of the rectus femoris muscle, an indirect tendon thickening that appears surrounded by an edematous area of signal hyperintensity (arrow). The hyperintensity area involves a tract extending in the proximal-distal direction for about 11 cm. The image is in agreement with a "bull eye lesion"

fibrotic tissue, must be interpreted as a possible risk factor for reinjury (Connell et al. 2004; Bedair et al. 2008; Gharaibeh et al. 2012).

4. However, attention must be paid to the fact that a hemosiderin deposit, consequent to bleeding, can mimic the formation of fibrotic tissue (Slavotinek 2010).
5. Given its greater sensitivity and the greater tissue contrast gradient, MRI is preferable to US when making RTT decisions (Slavotinek 2010; Connell et al. 2004).
6. The injured cross-sectional area derived from a fat-suppressed T2 in axial projection should be calculated as follows (Sanfilippo et al. 2013):

$$0.25\pi \text{ML} * \text{AP}$$

where ML is the medium-lateral distance and AP the anteroposterior distance of the hyperintensity area. The measurement must be carried out at the level of the area of greatest extension of the signal hyperintensity. It should be noted that it is also possible to calculate the volume of the entire hyperintensity zone (i.e., the volume of the injured zone) by adding the different cross-sectional areas and multiplying this value by the inter-slice distance.

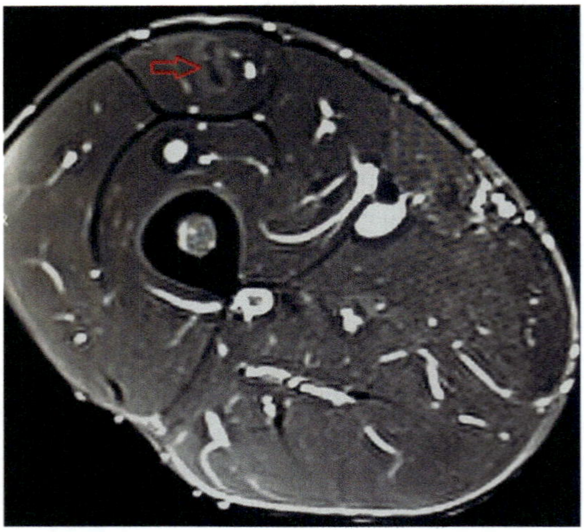

Fig. 5.2 The same injury in STIR axial MRI after 21 days. The edematous area appeared reduced by about 70%. The player entered, after the battery of laboratory and field tests, the RTT phase. During a follow-up of 6 months, he did not incur any reinjury

5.4 When Progress Unmasks the Dangerous Violation of the Wisdom of Nature

In sports traumatology, the latest few years have been characterized by significant progress in imaging. The two most used techniques in this area, namely, magnetic resonance imaging (MRI) and ultrasonography (US), have made remarkable progress in the latest decade. In the

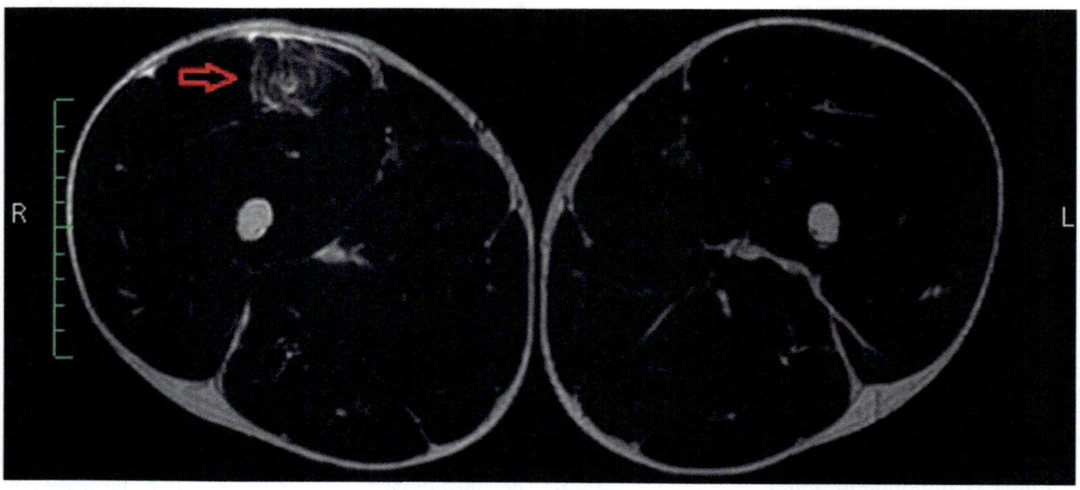

Fig. 5.3 MRI axial T2 showing "bull eye lesion" at indirect tendon of the right rectus femoris

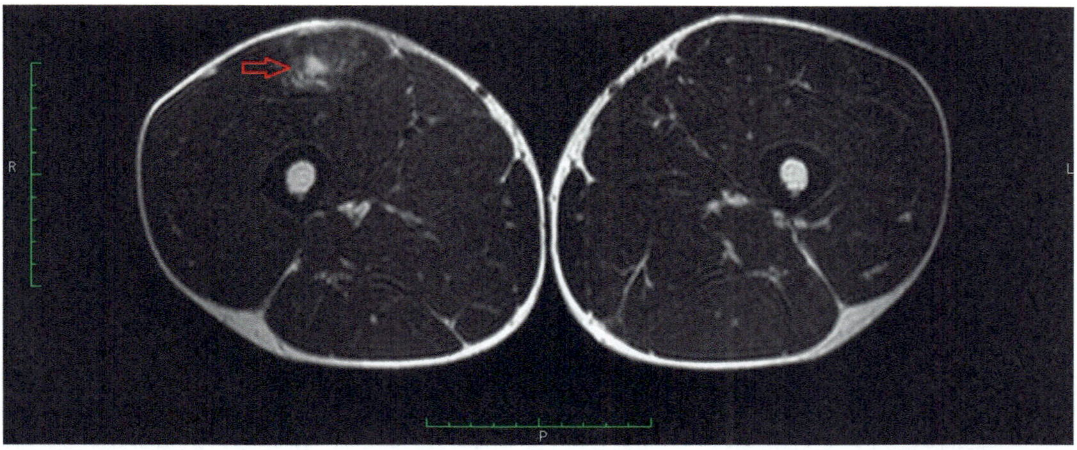

Fig. 5.4 MRI axial T2 of the same injury, performed after 25 days, showing a 70% decrease of the hyperintensity area with respect to the baseline. The player entered the RTT phase, and during a follow-up of 12 months, he did not incur any reinjury

study of muscle pathologies, modern MRI high field strength magnets are able to support a higher spatial and contrast resolution more easily. Therefore, it enables more accurate characterization of injuries of the muscular tissue, of the tendon, as well as those of the intramuscular connective tissue. Concerning MRI, some techniques such as diffusion-weighted imaging (DWI) and diffusion tensor imaging (DTI) seem particularly promising. DWI allows the measurement of molecular diffusion and is particularly interesting in the study of muscle inflammation. DTI allows the tracing of muscle fibers thus proving extremely suitable for the study of substructural muscle injuries. Even US, compared to its first clinical applications dating back to 1983, has made enormous technical progress, reaching a sensitivity of about 90% in the evaluation of muscle pathologies. However, for its greater spatial resolution and its greater contrast gradient, MRI represents the gold standard exam for the study of muscle-tendon pathologies. Obviously, this enormous improvement in imaging techniques should not exempt the clinician from having a high clinical capacity and a deep knowledge of semeiotics. Indeed, only a deep theoretical and practical knowledge of the latter, together with an equally high interpretive ability of imaging, will enable the formulation of a precise diagnosis. Recent advances in imaging have also allowed a follow-up of the muscle injury healing process, thus casting light on how some improvised therapeutic interventions can negatively affect the normal biological repair processes of muscle tissue. For example, interventions such as percutaneous fibrinolysis, deep massages, EWST, or early eccentric stimulation, performed on tissues still in the process of biological reparative reworking, can negatively interfere with the normal tissue reparative response, consequently altering the imaging control performed. The validity of those therapies used appropriately is beyond question; however, their use in excessively early stages on muscle tissue still in the process of repair/regeneration can be extremely counterproductive. Not only are the consequent imaging alterations extremely misleading for the clinician, especially in the event that the latter is unaware of the treatments causing such radiological alterations (as often happens in football, with many players embarking on therapeutic paths outside the club they belong to), they can also result in serious iatrogenic complications.

The assumption that the clinic is the queen in relationship to the imaging has always been so cited that it has essentially become a medical dogma, but this alleged dogmatism may, to date, not be justifiable in such situations.

Below are two practical examples concerning previous discussions in Figs. 5.5 and 5.6.

Our message intends to draw attention to the importance of carefully evaluating the therapies performed in sports medicine, since recent radiological techniques seem to show a higher-than-expected frequency of iatrogenic lesions.

In addition, we believe an important ethical and deontological reflection is worth attention.

Since the therapeutic treatments mentioned above are performed in an attempt (often vain) to accelerate the processes of tissue repair, we would like to provide the reader with food for thoughts with a quote Hans Jonas, an important philosopher and authoritative exponent of the current of Gnosticism, who in his book entitled "Technique, medicine and ethics" wrote: "Everyone must ask themselves whether it is good and wise, in consideration of the good of the individual and of the group, to bring confusion and ephemeral hedonism in the wisdom of nature which has established its times through a long evolution."

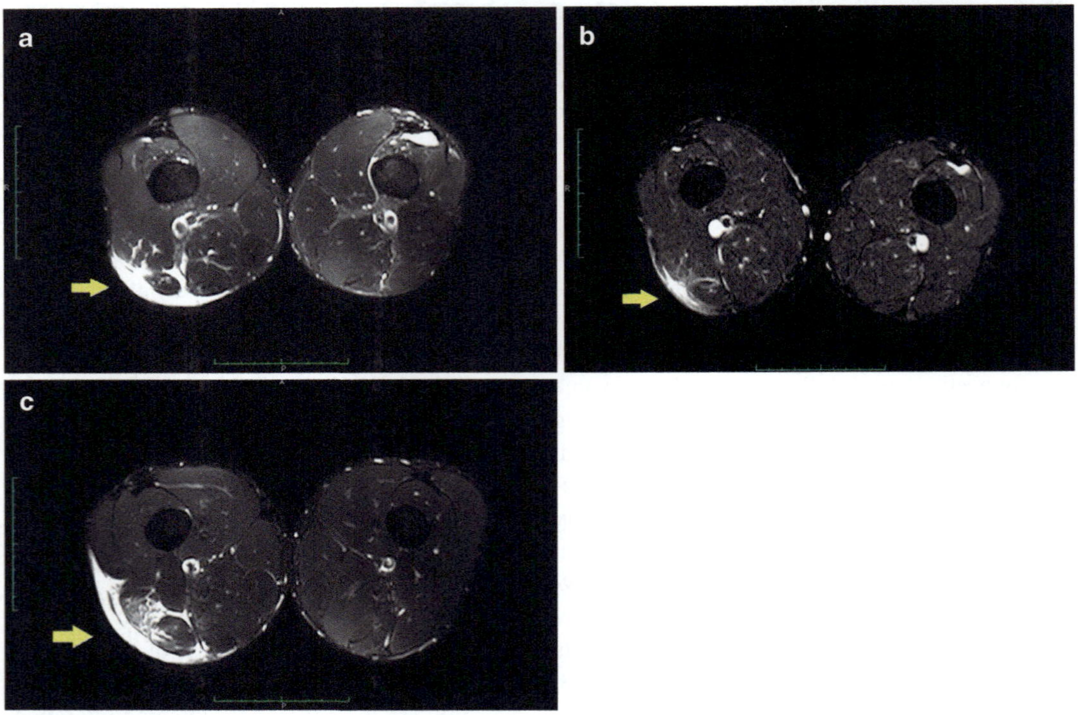

Fig. 5.5 (**a**) Axial MRI STIR image showing a second-degree injury of the right hamstring (arrow). (**b**) Axial STIR control image performed 12 days after the previous exam, showing a clear reduction of the signal hyperintensity (arrow). (**c**) Axial control STIR image performed 14 days after the second examination, showing a dramatic increase in the signal hyperintensity zone in the area undergoing biological repairs (arrow). This radiological worsening is probably due to a deep massage performed in the area still being repaired. It should be noted that no reinjury event had occurred during the rehabilitation process

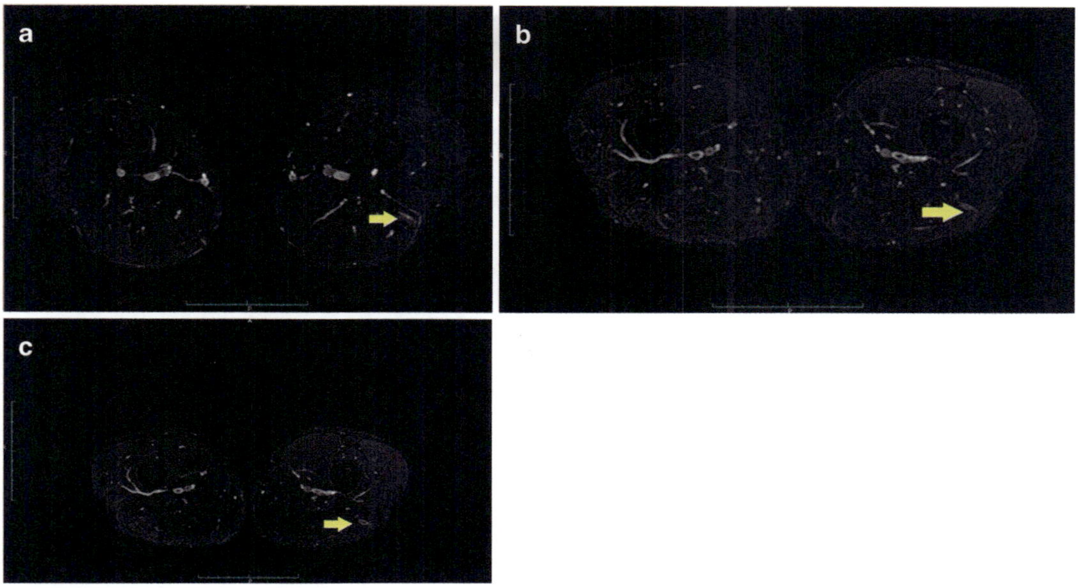

Fig. 5.6 (**a**) Axial T2 MRI image showing a first-degree lesion of the left hamstring. (**b**) Axial T2 control image performed 8 days after the previous examination showing an almost total disappearance of the hyperintensity zone. (**c**) T2 axial control image performed 10 days after the previous examination showing a reappearance of the signal hyperintensity in the regeneration zone (arrow). The radiological worsening is probably due to a percutaneous fibrinolysis operation performed in the area still undergoing biological repair. As in the previous case no reinjury event had occurred

5.5 Conclusions

Almost 90% of the clinically and functionally recovered muscle injuries showed intramuscular increased signal intensity on MRI fluid-sensitive sequences. The total normalization of the hyperintensity zone is not required for a successful and safe RTT. In the same time an important newly developed zone with fibrous tissues may be a reinjury risk factor. Furthermore, it is paramount to check that, during the therapeutic path, therapies that can negatively interfere with the natural biological repair process of muscle tissue are not performed.

References

Askling CM, Tengvar M, Saartok T, et al. Acute first-time hamstring strains during high-speed running: a longitudinal study including clinical and magnetic resonance imaging findings. Am J Sports Med. 2007;35:197–206.

Askling CM, Nilsson J, Thorstensson A. A new hamstring test to complement the common clinical examination before return to sport after injury. Knee Surg Sports Traumatol Arthrosc. 2010;18:1798–803.

Bedair HS, Karthikeyan T, Quintero A, et al. Angiotensin II receptor blockade administered after injury improves muscle regeneration and decreases fibrosis in normal skeletal muscle. Am J Sports Med. 2008;36:1548–54.

Brooks JHM, Fuller CW, Kemp SPT, et al. Incidence, risk, and prevention of hamstring muscle injuries in professional rugby union. Am J Sports Med. 2006;34:1297–306.

Comin J, Malliaras P, Baquie P, et al. Return to competitive play after hamstring injuries involving disruption of the central tendon. Am J Sports Med. 2013;41: 111–5.

Connell DA, Schneider-Kolsky ME, Hoving JL, Malara F, Buchbinder R, Koulouris G, Burke F, Bass C. Longitudinal study comparing sonographic and MRI assessments of acute and healing hamstring injuries. AJR Am J Roentgenol. 2004;183(4):975–84.

Gharaibeh B, Chun-Lansinger Y, Hagen T, et al. Biological approaches to improve skeletal muscle healing after injury and disease. Birth Defects Res C Embryo Today. 2012;96:82–94.

Heiderscheit BC, Sherry MA, Silder A, et al. Hamstring strain injuries: recommendations for diagnosis, rehabilitation, and injury prevention. J Orthop Sports Phys Ther. 2010;40:67–81.

Kerkhoffs GMMJ, van Es N, Wieldraaijer T, et al. Diagnosis and prognosis of acute hamstring injuries in athletes. Knee Surg Sports Traumatol Arthrosc. 2013;21:500–9.

Mendiguchia J, Brughelli M. A return-to-sport algorithm for acute hamstring injuries. Phys Ther Sport. 2011;12:2–14.

Reurink G, Goudswaard GJ, Tol JL, et al. MRI observations at return to play of clinically recovered hamstring injuries. Br J Sports Med. 2014;48(18):1370–6.

Reurink G, Brilman EG, de Vos R-J, et al. Magnetic resonance imaging in acute hamstring injury: can we provide a return to play prognosis? Sports Med. 2015a;45:133–46.

Reurink G, Almusa E, Goudswaard GJ, Tol JL, Bruce Hamilton B, et al. No association between fibrosis on magnetic resonance imaging at return to play and hamstring reinjury risk. Am J Sports Med. 2015b;43(5):1228–34.

Sanfilippo JL, Silder A, Sherry MA, Tuite MJ, Heiderscheit BC. Hamstring strength and morphology progression after return to sport from injury. Med Sci Sports Exerc. 2013;45(3):448–54.

Silder A, Heiderscheit BC, Thelen DG, et al. MR observations of long-term musculotendon remodeling following a hamstring strain injury. Skelet Radiol. 2008;37:1101–9.

Slavotinek JP. Muscle injury: the role of imaging in prognostic assignment and monitoring of muscle repair. Semin Musculoskelet Radiol. 2010;14:194–200.

6 Basic Principles of Dynamometric Test

6.1 Introduction

With the term dynamometry (from the Greek *dynamis* "strength" and *metron* "measure"), we mean the method by which muscle strength can be measured. The muscle strength can be expressed as (Barbat-Artigas et al. 2012):

1. Absolute value (kg, lbs)
2. In relationship with the considered lever arm (N/m)
3. In relationship with the body weight of the subject (N or kg/BW)
4. In relation to the lean mass of the subject (% of lean mass)
5. Muscle strength per unit of muscle mass ratio (MS/MM ratio)

Classically dynamometry can measure the production of force during all possible types of contraction for the human muscle, that is:

1. Isometric contraction
2. Concentric contraction
3. Stretch-shortening cycle
4. Eccentric contraction
5. Isokinetic contraction

Each different type of contraction listed above has its biomechanical and biological specificities, which is why they all provide different types of information concerning the muscle behavior.

6.2 The Dynamometric Assessment of the Isometric Contraction

A muscle contraction is defined as isometric when the muscle is stimulated at a constant length, without allowing its ends to approach (Hakkinen 1993; Baker et al. 1994; Murphy et al. 1995).

Experiments involving isometric contractions are carried out in the laboratory, on isolated muscle, in order to determine the strength developed by the muscle in function of the time. Furthermore, the isometric evaluation of muscle function is a practice widely used also in muscle in vivo, and it is today considered a valid investigation method to study the biomechanical characteristics of skeletal muscle (Murphy et al. 1995; Bisciotti 1998). In general, these types of tests are used to quantify both the isometric maximum force and its production as a function of time (RFD, Rate of Force Development). However, over the years, the ability of isometric strength tests to provide valid indications relating to the dynamic behavior of the muscle has been gradually questioned (Wilson et al. 1991; Wilson and Murphy 1997). Indeed, if we examine the correlation between the maximum isometric force and dynamic performance, we cannot help expressing some doubts about the former's validity as an indicator of the dynamic function of the muscle. In fact, even if some authors demonstrate valid correlation with sports

G. N. Bisciotti et al., *Return-to-Play after Lower Limb Muscle Injury in Football*,
https://doi.org/10.1007/978-3-030-84950-4_6

performance, as in the case of basketball (Hakkinen 1987), rowing (Secher 1975), and the fast disciplines of athletics (Mero et al. 1981), many other studies underline the minimum correlation existing between the maximum isometric force and dynamic performance. Lack of correlation is shown even in the case of very specific test protocols concerning the investigated sport discipline (Considine and Sulliwan 1973; Viitasalo et al. 1981; Jaric et al. 1989; Bloomfield et al. 1990; Strass 1991; Murphy et al. 1994). For this reason, although the isometric maximal force can be considered as an index of the maximum strength production, the indications that can be obtained from the latter regarding the dynamic behavior of the muscle are at least questionable (Atha 1981; Enoka 1994; Wilson et al. 1995). Furthermore, it is necessary to remember that isometric tests are poorly sensitive to muscle variations induced by training carried out in a dynamic regime (Baker et al. 1994; Wilson and Murphy 1997). A further use of isometric force tests is made to quantify the rate of force development of the isometric force (RFD). For this purpose, different test protocols are used. Indeed, even if in general, the RFD is interpreted as the point of maximum slope in the force-time curve, several authors identify different calculation methods for its quantification. The RFD calculation interval can range from 5 to 100 ms (Viitasalo et al. 1980; Bürle 1985; Golhofer 1987; Baker et al. 1994; Christ et al. 1994; Wilson et al. 1994, Verchoshansky 1996.). Other authors use a quantification of the RFD based on the time required to reach a certain percentage of the maximal isometric force (Hakkinen et al. 1985; Viitasalo et al. 1980) or calculate the integral of the force over the time on different intervals of this latter (Verchoshansky 1996). However, the common basis of all RFD quantification protocols is the investigation of the ability to develop an appreciable percentage of the maximum isometric force in a short time. For this reason, RFD is interpreted by many authors as an isometric index that could be better related to dynamic performance. However, even in this field of investigation, the results are contradictory. Indeed, some authors (Viitasalo et al. 1981) have shown how RFD is related, albeit modestly (0.49–0.66), with the execution of ballistic movements of the lower limbs. In other studies (Viitasalo and Aura 1984), the isometric RFD was much more closely related to the jump performance ($r = 0.9$), or it turned out to be a more important discriminating parameter for dynamic performance, compared to the simple value of isometric maximal force (Komi 1984). However, numerous other authors have found no significant correlation between isometric RFD and dynamic performance (Mero et al. 1981; Wilson et al. 1995; Young and Bilby 1993; Pryor et al. 1994). The reasons for this poor correlation between the different isometric strength indices and muscular behavior in dynamic conditions, which was pointed out by many authors, can be found in the considerable differences in the nervous and mechanical behavior between isometric and dynamic muscle activation (Wilson et al. 1995). Indeed, at the basis of the difference between dynamic and isometric activation of the muscle, neurogenic factors inherent in the type of recruitment may reside. It is also important to remember that one of the most important limitations of the isometric evaluation is the specificity of the force values related to the joint angle. The isometric force value is strongly linked to the joint angle of measurement, due to the fact that to different joint angles correspond different degrees of covering of the actin and myosin myofilaments and therefore different strength values (Cavagna 1988; Sale 1991; Duchateau and Hainaut 1984). The data above represent one of the most important problem concerning the measurement of the maximal isometric force and, consequently, of the RFD. Indeed, it cannot be assumed that valid correlations exist between the isometric force values obtained at the different joint angles in a given movement cycle (Wilson et al. 1995). In the literature, at our knowledge, there is no research showing that every joint angle of the movement is representative of the totality of the movement itself. In this regard, a study of Murphy et al. (1995) shows the relationship between the maximal isometric force values recorded during the execution of the bench press exercise at elbow joint angles of 90° and 120°. The study shows that the correlation between the force values measured at the two joint angles is 0.7, thus indicating that more than 50% of the variance was not common at the two considered joint angles. These data agree with those showed from another similar experience, where the maximal isometric force values for the extensor muscles of the upper limbs during the bench press exercise, recorded respectively at 25% of the entire movement cycle (equal at an elbow joint angle of 80 ± 10°) and at 75% (equal to an elbow joint angle of 116.3 ± 12.3°) had a correlation of 0.74 (Bisciotti and Belli 1997, unpublished study). Furthermore, Murphy and Wilson (1996) have highlighted, using spectrum analysis, how isometric and dynamic activations are characterized both by different motor unit recruitment patterns and by different activation frequencies. These considerations support the hypothesis of different types of neuromuscular activation between the two mechanisms, which was already advanced by Henry and Whitley (1960), although their conclusions were not at the time supported by electromyographic investigations. Indeed, the motor units could follow the classic principle of recruitment of Hennemann et al. (1965) only in particular conditions of isometric activation, while dynamic activation could involve a very different recruiting pattern (Person 1974; Caldwell et al. 1993; Nakazawa et al. 1993). In this regard, it should be noted that the electromyography activity is considerably higher in the case of a ballistic movement compared to isometric muscle activation, emphasizing the difference in muscle recruitment patterns (Desmet and Godaux 1997). The difference in the activation pattern between isometric and dynamic contraction is also underlined by several other authors (Person 1974; Nakazawa et al. 1993). These authors agree that in the case of isometric contractions, both the fast-twitch fibers and the slow-twitch fibers contribute to the development of force, while, in the case of dynamic ballistic movements, the fast-twitch fibers seem to contribute more in the development of muscle tension than the slow-twitch fibers. This theory is also supported by the fact that the electromyographic activity recorded during the execution of a ballistic movement has significantly higher values than those of an isometric contraction (Nakazawa et al. 1993). Furthermore, in the isometric contraction, the increase in muscle tension occurs mainly through the recruitment of new motor units: a linear correlation between the muscle tension increase and the increase in the electromyographic signal can be observed (Nakazawa et al. 1993). On the contrary, in a dynamic activation, the electrical activity experiences a statistically significant increase when going from 40% to 50% of the maximum load. Above this limit, it is possible to notice a flattening of the electrical activity. Such data demonstrate that all possible motor units have already been recruited at 50% of the maximum load and a further increase in muscle tension can only occur through an increase in the discharge frequency (Wilson et al. 1994). Another aspect of mechanical nature, which can differentiate isometric contraction from dynamic contraction, is the muscle-tendon stiffness. The stiffness of the muscle-tendon unit appears to be more related to isometric activation than to dynamic activation (Wilson et al. 1994). Conceptually, a more rigid muscle-tendon unit should increase the production of force in both isometric and dynamic conditions. However, it should be emphasized that during the dynamic contraction the contractile component is shortened mainly due to the movement in progress, while in the isometric contraction the shortening of the contractile component is mainly due to the lengthening of the muscle-tendon unit. For this reason, a greater muscle-tendon stiffness would be more important in the isometric contraction than in the concentric one (Wilson et al. 1994). Lastly, it should be remembered that the dynamic performances involving a stretch-shortening cycle entail, unlike in the isometric contraction, a considerable use of both elastic energy (Wilson and Murphy 1997) and residual force enhancement (RFE) (Seiberl et al. 2015). The RFE phenomenon can be described as the increase in strength that can be made in steady-state isometric forces following an active muscle stretch (Seiberl et al. 2015). The RFE underlying mechanisms are still not entirely understood (Herzog 2014), but the phenomenon was observed at all muscle lengths (Rassier 2012; Peterson et al. 2004) and is independent of stretch velocity (Edman et al. 1982; Sugi and Tsuchiya 1988).

For all the reasons listed above, in our opinion, the data recorded from the isometric tests could provide false information if they were used as forecast indicators for dynamic performance. Consequently, we suggest caution in their interpretation in the rehabilitation process control. Indeed, in light of the evidence found in the literature, we must recognize the superiority of dynamic tests in the study of muscle function compared to isometric tests. Such superiority is due to dynamic tests allegedly involving activation mechanisms that are essentially analogous to the dynamic performance investigated. However, isometric tests are particularly suitable during the early stages of the rehabilitation process, when the range of joint motion is reduced and muscle function is still low. The basic technical principles to be observed during the execution of an isometric test are summarized in Table 6.1.

6.3 Isotonic Assessment Versus Isokinetic Assessment

The isokinetic methodology has undoubtedly allowed important progress in the study of muscle behavior (Hislop and Perrine 1967; Perrine and Edgerton 1968). However, the type of muscle contraction produced through the isokinetic modality presents significant and essential differences with the type of contraction that is carried out during a natural movement. The first fundamental difference between isokinetic and natural contraction is the fact that, in the first case, the muscle contracts at constant speed, without therefore being able to generate acceleration, which on the contrary is one of the main characteristics of natural contraction. Secondly, it should be remembered that most natural movements, both in humans and in animals, are characterized by muscular activation which involves an eccentric muscle contraction phase, followed by a very limited isometric stabilization phase, immediately followed by a concentric phase. This pattern of activation is called "stretch-shortening cycle" (SSC) (Norman and Komi 1979; Goubel 1987; Komi 1987; Seiberl et al. 2015).

During a SSC, the force production is greater than the force produced during a pure concentric movement (i.e., a movement that excludes the eccentric phase) for the following reasons:

1. The dynamics activation: the stretch induced increased contractility (Seiberl et al. 2015).
2. The contributions of stretch reflexes (Ettema et al. 1992; Walshe et al. 1998).
3. The storage and release of elastic energy (van Ingen Schenau et al. 1997a, b).
4. The contribution of the residual force enhancement phenomenon (Seiberl et al. 2015).

Basically, from a pre-stretching phase immediately followed by a concentric contraction, there is an accumulation of potential elastic energy for the muscle-tendon complex, which is returned in the form of mechanical work, during the concentric phase of the movement (Goubel 1987). This represents a first important difference between a natural movement performed through SSC and an isokinetic movement. Indeed, during an isokinetic exercise or test, the accumulation of elastic energy during the eccentric phase of the movement is prevented by the resistance offered by the device, which is proportional to the force expressed by the subject. This means that the

Table 6.1 Basic principles for the administration of dynamometric (isometric, isotonic, and isokinetic) tests

Isometric test (Webber and Porter 2010; Toonstra and Mattacola 2013; Ruschel et al. 2015)	Isotonic tests (Webber and Porter 2010; Bieler et al. 2017; Van Driessche et al. 2018a, b)	Isokinetic tests (Croisier et al. 2002; Orchard et al. 2005; Sanfilippo et al. 2013; Menzel et al. 2013; Bisciotti et al. 2016)
Operate a proper warm-up	Operate a proper warm-up	Operate a proper warm-up
Biomechanically isolate the muscle group to be tested	Biomechanically isolate the muscle group to be tested	Biomechanically isolate the muscle group to be tested
Standardize the lever arm	Standardize the lever arm and ROM	Standardize the lever arm and ROM
Begin the test with the healthy limb	Begin the test with the healthy limb	Begin the test with the healthy limb
Apply an isometric contraction of progressive intensity for a duration of between 3 and 5 s	Apply the maximal speed during the movement	Align the center of rotation of the joint with the center of rotation of the mechanical device
Encourage the patient during the test	Encourage the patient during the test	Encourage the patient during the test
Perform at least three trials with an adequate recovery between each trial (around 1 min 30 s)	Perform at least one set of six to ten repetitions	Subtract the weight of the limb from the calculation of the force moment (usually done automatically by the device)
Consider the peak value	Consider both average and peak value	Consider both average and peak value, to avoid the so-called "peak artifact"
Check for any pain symptoms with VAS	Check for any pain symptoms with VAS	Check for any pain symptoms with VAS
Stop the test in the presence of severe pain (VAS >3)	Stop the test in the presence of severe pain (VAS >3)	Stop the test in the presence of severe pain (VAS >3)
The dynamometric values must be ≥90% of the pre-injury values or ≥90% of the contralateral limb values	The dynamometric values must be ≥90% of the pre-injury values or ≥90% of the contralateral limb values	Perform one set of 6–10 repetitions at low speed (30°/s–60°/s) and one set at high speed (>300°/s)
		Perform at least one eccentric test at 60°/s or 30°/s
		Consider the value of the joint angle corresponding to the peak force production
		Consider the values of the mechanical work
		Consider the shape of the force curve
		Consider the value of the ratio of HS (concentric modality) to Q (concentric modality), and the value of the ratio HS (eccentric modality) to Q (concentric modality)
		Perform the tests observing an adequate recovery between the sets (~2–3 min)
		The dynamometric values must be ≥90% of the pre-injury values or ≥90% of the contralateral limb values

subject does not need to exert any eccentric force to break the load, which will not tend to return to the starting position due to the gravity effect. Furthermore, during a natural SSC, the muscle is able to develop very relevant angular speeds, which in the case of the elbow joint can even exceed 34 radiants · s^{-1} (i.e., 1948°/s) (Respizzi 1997). On the contrary, the angular speeds that can be reached during a movement carried out on an isokinetic equipment can be maximum of six to seven radiants · s^{-1} (i.e., 343–401°/s). For a better understanding, below are some examples of speed recorded during various activities at the level of the knee joint:

- Walking: 233°/s
- Running: 1200°/s
- Punting: 2865°/s

Some other estimated speed values during sport movements are:

- The shoulder during throwing: ≥7000°/s
- The elbow during throwing: ≥1825°/s
- The ankle during running: ≥540°/s

In addition to that, there are also marked differences in the electromyographic pattern recorded during an isokinetic in comparison to a natural contraction (also definable as heterotonic contraction) (Hislop and Perrine 1967). Further points concerning the isokinetic condition should be underlined:

1. The isokinetic condition, that is, the respect of constant speed, is not always observed during an isokinetic movement. Actually, during the initial phase, which is called "oscillatory phase," the isokinetic equipment requires a certain interval of time to adjust the set speed. During this interval the movement does not take place at a constant speed (Fig. 6.1). Furthermore, during the oscillatory phase, the higher the set speed, the longer it takes for the equipment to control the set speed. Therefore, if the set speed is relatively high, the speed actually reached during the first phase of the movement can be even 50% higher than the programed one (Gransberg and Knutsson 1983). The isokinetic condition is not respected even in the final phase of the movement (Fig. 6.1), called "deceleration phase" (Gransberg and Knutsson 1983). For this reason, in a rather limited range of movement, such as a leg extension movement carried out at the isokinetic device (whose angular range is only 90°), what can actually be defined as isokinetic is only the central part of the movement (Respizzi 1997). Therefore, we are faced with a movement carried out in a "hybrid" and not well-defined form, which constitutes a "mix" between constant and variable speed.
2. The power production, during an isokinetic test, is systematically underestimated. As well known in the force-speed relationship concerning the muscle in vitro, the maximum power is obtained with the production of a force representing about 1/3 of the maximum force of the subject and a shortening speed which is equal to 1/3 of the maximum speed of contraction (Hill 1938). On the contrary, in the muscle in vivo (i.e., in conditions of natural activation), the force-speed relationship is linear, and the peak power is obtained through a force production and a contraction speed both equal to 50% of their maximum

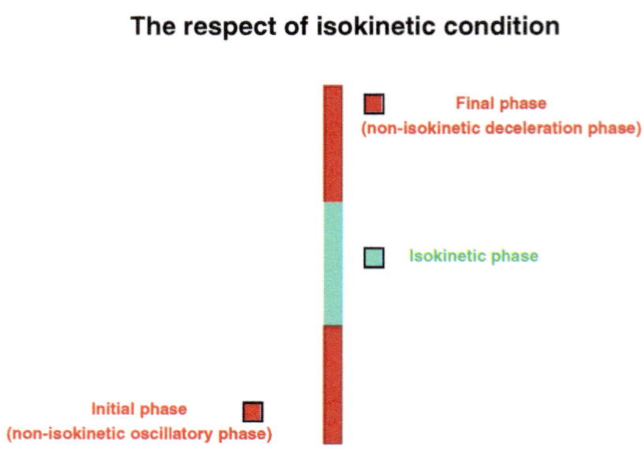

Fig. 6.1 The isokinetic condition is not always observed during an isokinetic movement. As shown in figure, during the "oscillatory phase," the isokinetic equipment requires a certain interval of time to adjust the set speed. In this time span, the movement is not isokinetic. Even in the final deceleration phase, the movement does not take place at constant speed. Therefore, the true isokinetic phase is reduced only to the central phase of the movement (Gransberg and Knutsson 1983)

value (Catlaw et al. 1996; Driss et al. 1998; Jaskólska et al. 1999; Bisciotti et al. 2000). The problem is that the best isokinetic dynamometers produce a maximum angular speeds between about 400°/s and 450°/s, compared to maximum speeds during a natural movement that are greater than 1000°/s, as in the case of the shoulder joint. For this reason, it is easily understood that, during an isokinetic evaluation, the maximum power of the tested muscle group is systematically underestimated.

3. The moment in which the peak of force is recorded in an isokinetic movement totally differs from the moment in which the maximum expression of force occurs during a heterotonic movement. During a heterotonic exercise (better defined with the term "auxotonic," since different strength values correspond to different angles of movement) the movement of the bone levers takes place at a variable speed and constant load. During this type of movement, the variation of the joint lever arm corresponds to a variation in the muscle strength production. For this reason, the maximum force value will be recorded at the most unfavorable moment of the joint lever arm. In isokinetic devices, muscle strength is calculated as a "force moment" and therefore corresponds to the product between the force and the distance between the point of application of the force and the center of rotation [Force moment (Nm) = force X lever arm]. Therefore, during an isokinetic movement, the peak of the force moment (also called peak torque) is recorded at the most favorable point of the joint lever arm (Respizzi 1997).

However, isokinetic devices show two important advantages:

1. They have excellent reproducibility of the values recorded during the tests.
2. Many types of isokinetic equipment allow the recording of the eccentric force values of the tested muscles. This type of investigation would be particularly interesting especially for those muscles, such as the hamstring, which are particularly stressed in eccentric contraction during the natural movements. However, the eccentric force value that is recorded refers to an eccentric movement carried out at constant speed (taking into account the limits discussed above regarding the constant speed during an isokinetic movement), which differs from that occurring during a natural eccentric movement which, on the contrary, occurs at variable muscle stretching speed.

For all this series of reasons, the values derived from the isokinetic tests must be interpreted with caution. However, the isokinetic method, despite its limitations, represents a valid instrument of investigation of muscle function in the athlete.

6.4 The Interpretation of the Isokinetic Test

From the information gathered during the isokinetic tests several important values can be obtained:

1. Torque parameters
2. Time rate to torque development
3. Force decay rate
4. Range of motion
5. Muscular performance parameters

6.4.1 Torque Parameters

The peak torque (PT) represents the highest point on the extension/flexion curve. On the contrary, the mean peak torque (MPT) is the average of the peak torque value recorded during a series of repetitions. The MPT allows a better estimation of the overall function when compared to PT, since the muscular function is dependent on the repetitions of the movement.

6.4.2 The Time Rate to Torque Development

The time rate to torque development (TRTD) is the equivalent of the rate of force development during the isometric tests (Figs. 6.2 and 6.3). It indicates how rapidly the peak torque is reached during the isometric contraction. Normally, in a healthy subject, the peak torque is reached during the first one-third of the upward slope of the torque curve. If the upward slope of the curve is prolonged and the peak torque occurs in the middle or in the final third of the curve, the subject has difficulty in generating the torque at the onset of the muscle contraction. Such situation is indicative for a muscular functional deficit. The TRTD in isokinetic conditions may be calculated and examined in relation to the following parameters:

1. The peak torque measured from the start of the muscle contraction to the highest point on the torque curve.
2. A predetermined torque value (i.e., the time necessary to produce a specific torque value is calculated).
3. A predetermined time (i.e., the torque produced in a specific time is calculated).
4. A specific angle (i.e., the torque produced at a specific angle is calculated).

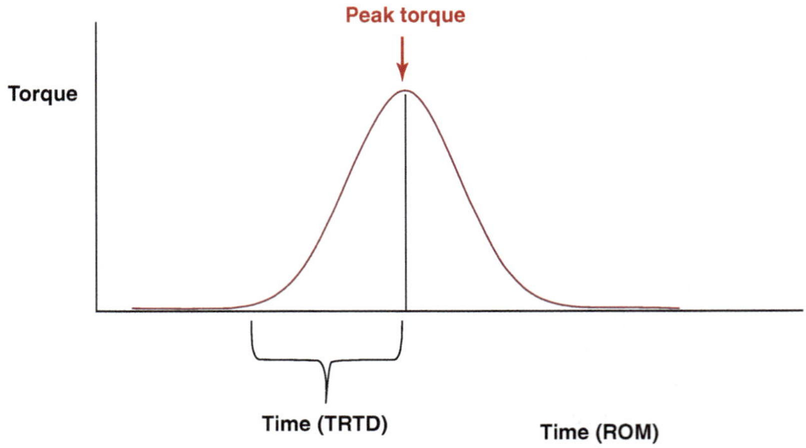

Fig. 6.2 The bell-shaped torque curve. On the abscissa axis the time and in the ordinate axis the force. The peak torque value is represented in function of the time

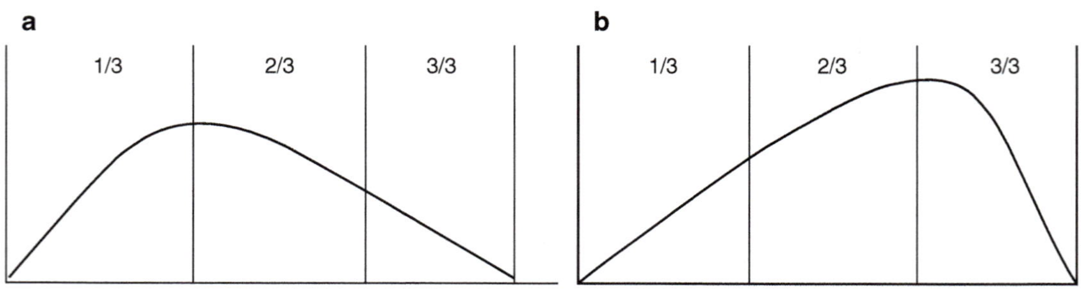

Fig. 6.3 A normal curve (**a**) showing the peak torque occurring during the first one-third of upward slope. In (**b**) a prolonged TRTD indicates difficulty generating muscle contraction

Depending upon the pathologies, certain numbers of different parameters may be taken into account. For example, knowing that by 0.2 s, the knee joint should be able to achieve 80–90% of its torque, the knee angle at heel strike is between 0° and 5°, and the knee angle at mid-stance is ~20° (Wilk et al. 1991), we know that an important value of torque is needed for dynamic stability of the knee during walking. Indeed, concerning the knee joint evaluation, TRTD represents a very useful biomechanical parameter to be collect during an isokinetic test (Wilk et al. 1991).

6.4.3 Force Decay Rate

The force decay rate (FDR) represents the downslope of a torque curve. FDR is the section of the curve showing where torque value decreases. Usually, the downslope of the torque curve should be either straight or convex (Fig. 6.4). A concave downslope may indicate that the subject has difficulty in the force production as he/she reaches the terminal extension. These data may be indicative for some pathologies. For example, in subjects with ACL, deficit concavity tends to occur between 20° and 30° (Wilk et al. 1991).

6.4.4 Range of Motion

The drop in strength to some degree of the ROM indicates the precise point where muscular deficiencies occurs. The ROM evaluation is also a useful indicator for test validity and reliability. The torque peak should be produced, both in injured and non-injured limbs, approximately at the same articular angle both during a concentric contraction and during an eccentric contraction (Fig. 6.4). That is particularly important during eccentric contraction. Indeed, an eccentric peak torque produced in the injured limb at a lower angle in comparison to the non-injured limb (Fig. 6.5), may show a difficulty for the injured muscle to produce eccentric force at important muscle lengths. This may represent an important risk factor for reinjuries (Bisciotti et al. 2019).

6.4.5 Reciprocal Innervation Time

The reciprocal innervation time (RIT) represents the amount of time elapsing between the end of agonistic contraction and the onset of antagonistic contraction. Generally, the RIT is useful to study how quickly a subject is able to begin

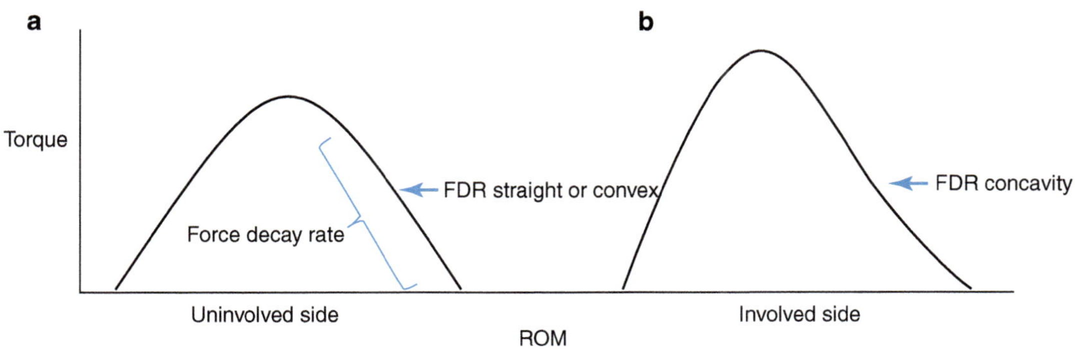

Fig. 6.4 (**a**) Shows a normal FDR pattern, while (**b**) pictures a concave downslope indicating difficulty in force production

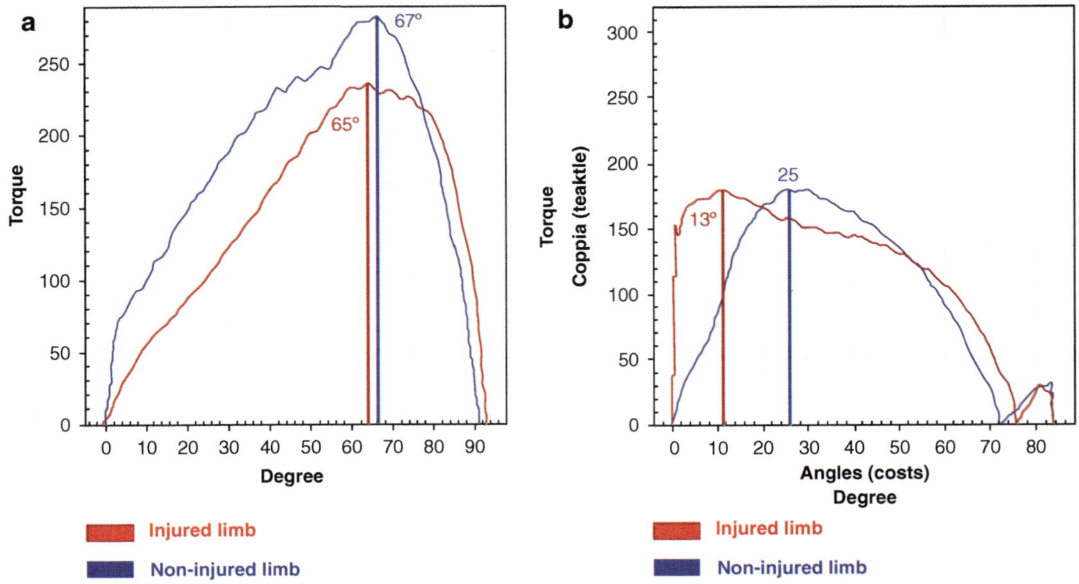

Fig. 6.5 In (**a**) the peak torque is produced substantially at the same joint angle both for injured and non-injured limb: On the contrary, in (**b**), the peak torque is produced at very different joint angles. This may represent an important risk factor for reinjuries

the hamstring contraction following a quadriceps contraction. The portion of a torque curve between the end of the extension and the beginning of the flexion (and vice versa) appears as a “U” rather than a “V” if there is a delay in RIT.

References

Atha J. Strengthening muscle. In: Miller DI, editor. Exercise and sport science reviews, vol. 9. Philadelphia: Franklin Institute Press; 1981. p. 1–73.

Baker D, Wilson G, Carlyon B. Generality versus specificity: a comparison of dynamic and isometric measures of strength and speed-strength. Eur J Appl Physiol. 1994;68:350–5.

Barbat-Artigas S, Rolland Y, Zamboni M, Aubertin-Leheudre M. How to assess functional status: a new muscle quality index. J Nutr Health Aging. 2012;16(1):67–77.

Bieler T, Magnusson SP, Christensen HE, et al. Muscle power is an important measure to detect deficits in muscle function in hip osteoarthritis: a cross-sectional study. Disabil Rehabil. 2017;39:1414–21.

Bisciotti GN. I valori di forza isometrica possono costituire un valore predittivo della performance dinamica? Coach Sport Sci J. 1998;3(1):47–58.

Bisciotti GN, Combi F, Forloni F. Per ritrovare la funzionalità. Sport Med. 2000;6:43–7.

Bisciotti GN, Quaglia A, Belli A, et al. Return to sports after ACL reconstruction: a new functional test protocol. Muscl Ligament Tend J. 2016;06:499–509.

Bisciotti GN, Volpi P, Alberti G, Aprato A, Artina M. Auci A et al. BMJ Open Sport Exerc Med. 2019;5(1):e000505. https://doi.org/10.1136/bmjsem-2018-000505.

Bloomfield J, Blanksby BA, Ackland TR, Allison GT. The influence of strength training on overhead throwing velocity of elite water polo players. Aust J Sci Med Sport. 1990;22:63–7.

Bürle M. Dimensionen des Kraftverhaltens und hire spezifischen Trainings methoden. In: Bürle M, editor. Grundlagen des Maximal und Schnellkrafttraining, vol. 3. Schorndorf; 1985. p. 82–111.

Caldwell G, Jamison J, Lee S. Amplitude and frequency measures of surface electromyography during dual task elbow torque production. Eur J Appl Phys. 1993;66:349–56.

Catlaw K, Arnold BL, Perrin DH. Effect of cold treatment on the concentric and eccentric torque-velocity relationship of the quadriceps femoris. Isokinet Exerc Sci (Stoneham, Mass). 1996;3:157–60.

Cavagna G. Muscolo e locomozione. Cortina Editions Milano; 1988.

Christ CB, Slaughter MH, Stillman RJ, Cameron J, Boileau RA. Reliability of selected parameters of isometric muscle function associated with testing 3 days X 3 trials in women. Strength Cond Res. 1994;8:65–71.

Considine W, Sulliwan W. Relationship of selected tests of leg strength and leg power on college man. Res Q. 1973;44:404–15.

Croisier JL, Forthomme B, Namurois MH, et al. Hamstring muscle strain recurrence and strength performance disorders. Am J Sports Med. 2002;30:199–203.
Desmet J, Godaux L. Ballistic contraction in man: characteristic recruitment pattern of single motor units of the tibialis anterior muscle. J Phisiol. 1997;264:673–93.
Driss T, Vandewalle H, Monod H. Maximal power and force-velocity relationship during cycling and cranking exercises in volleyball players. Correlation with the vertical jump test. J Sports Med Phys Fitness. 1998;4(38):286–93.
Duchateau JK, Hainaut K. Isometric or dynamic training: differential effects on mechanical properties of human muscle. J Appl Physiol. 1984;56:296–301.
Edman KA, Elzinga G, Noble MI. Residual force enhancement after stretch of contracting frog single muscle fibers. J Gen Physiol. 1982;80:769–84.
Enoka RM. Neuromechanical basis of kinesyology. 2nd ed. Champaign. IL: Human Kinetics; 1994.
Ettema GJ, Huijing PA, de Haan A. The potentiating effect of prestretch on the contractile performance of rat gastrocnemius medialis muscle during subsequent shortening and isometric contractions. J Exp Biol. 1992;165:121–36.
Golhofer A. Komponenten der Schnellkraftleistungen im Dehnungs. Erlensee: Verkurzungszyklus; 1987.
Goubel F. Muscle mechanics. Med Sport Sci Series. 1987;26:24–35.
Gransberg L, Knutsson E. Determination of the dynamic muscle strength in man with acceleration controlled isokinetic movements Acta Physiol. Scandinavica. 1983;119:317–20.
Hakkinen K. Force production characteristics of leg extensor, trunk flexor, and extensor muscles in male and female basketball players. J Sport Med Phis Fit. 1987;31:325–31.
Hakkinen K. Neuromuscular fatigue and recovery in male and female athletes during heavy resistance exercise. Int J Sport Med. 1993;14:53–9.
Hakkinen K, Komi PV, Alen M. Effect of explosive type strength training on isometric force and relaxation-time, electromyographic and muscle fibre characteristics of leg extensor muscle. Acta Physiol Scand. 1985;125:587–600.
Hennemann E, Somjen G, Carpenter DO. Functional significance of cell size in spinal motoneurons. J Neurophysiol. 1965;28:555–60.
Henry FM, Whitley JD. Relationship between individual differences in strength speed, and mass of an arm movement. Res Q. 1960;31:24–33.
Herzog W. Mechanisms of enhanced force production in lengthening (eccentric) muscle contractions. J Appl Physiol. 2014;116:1407–17.
Hill AV. The heat of shortening and the dynamic constant of muscle. Proc Roy Soc B. 1938;126:136–95.
Hislop HJ, Perrine JJ. Isokinetic concept of exercise. Phis Ter. 1967;47:114–7.
Jaric S, Ristanovic D, Corcos DM. The relationship between muscles kinematic parameters and kinematic variables in a complex movement. Eur J Appl Physiol. 1989;59:370–6.
Jaskólska A, Goossens P, Veenstra B, Jaskólski A, Skinner JS. Comparison of treadmill and cycle ergometer measurements of force-velocity relationship and power output. Int. J. Sports Med. 1999;20:192–7.
Komi PV. Physiological and biomechanical correlates of muscle function: effects of muscle structure and stretch-shortening cycle on force and speed. In: Terjung RL, editor. Exercise and sport science reviews, vol. 12. Toronto: D.C. Heat; 1984. p. 81–122.
Komi PV. Elastic potentation of muscle and its influence on sport performance. In: Bauman W, editor. Biomechanics and performance in sport; 1987. p. 59–70.
Menzel HJ, Chagas MH, Szmuchrowski LA, et al. Analysis of lower limb asymmetries by isokinetic and vertical jump tests in soccer players. J Strength Cond Res. 2013;27:1370–7.
Mero A, Luhtanen P, Viitasalo JT, Komi PV. Relationship between maximal running velocity, muscle fiber characteristics, force production and force relaxation of sprinters. Scand Sport Sci. 1981;3:16–22.
Murphy AJ, Wilson GJ. Poor correlation's between isometric tests and dynamic performance: relation to muscle activation. Eur J Appl Physiol. 1996;73:353–7.
Murphy AJ, Wilson GJ, Pryor JF. The use of the isoinertial force mass relationship in the prediction of dynamic human performance. Eur J Appl Physiol. 1994;69:250–7.
Murphy AJ, Wilson GJ, Pryor JF, Newton RU. Isometric assessment of muscular function: the effect of joint angle. J Appl Biomech. 1995;11:205–15.
Nakazawa K, Kawakami Y, Fukunaga T, Yano H, Miyashita M. Differences in activation patterns in elbow flexor muscles during isometric, concentric, and eccentric contraction. Eur J Appl Physiol. 1993;66:214–20.
Norman RW, Komi PV. Electromyographyc delay in skeletal muscle under normal movement condition. Acta Physiol Scand. 1979;106:241.
Orchard J, Best TM, Verrall GM. Return to play following muscle strains. Clin J Sport Med. 2005;15:436–41.
Perrine JJ, Edgerton VR. Muscle force-velocity and power-velocity relationship under isokinetic loading. Med Sci Sport. 1968;10:159–66.
Person R. Rhythmic activity of a group of human motoneurons during voluntary contraction of a muscle. Electroencephalogr Clin Neurophysiol. 1974;36: 585–95.
Peterson DR, Rassier DE, Herzog W. Force enhancement in single skeletal muscle fibres on the ascending limb of the force-length relationship. J Exp Biol. 2004;207:2787–91.
Pryor JF, Wilson GJ, Murphy AJ. The effectiveness of eccentric, concentric and isometric rate of force development tests. J. Hum Mov Stud. 1994;27:153–72.
Rassier DE. The mechanisms of the residual force enhancement after stretch of skeletal muscle: non-uniformity in half-sarcomeres and stiffness of titin. Proc Biol Sci. 2012;279:2705–13.

Respizzi S. L'esercizio isocinetico: applicazioni e limiti in riabilitazione sportiva. In: Aggiornamenti in riabilitazione sportiva. Milano: Edi-ermes Ed; 1997.

Ruschel C, Haupenthal A, Jacomel GF, Fernandes Jacomel G, et al. Validity and reliability of an instrumented leg-extension machine for measuring isometric muscle strength of the knee extensors. J Sport Rehabil. 2015;24:2013–122.

Sale DG. Testing strength and power. In: Mac Dougall J, Wenger H, Green H, editors. Physiological testing of the high performance athlete. 2nd ed. Champaign, IL; 1991. p. 21–106.

Sanfilippo JL, Silder A, Sherry MA, et al. Hamstring strength and morphology progression after return to sport from injury. Med Sci Sports Exerc. 2013;45: 448–54.

Secher NH. Isometric rowing strength of experienced and inexperienced oarsman. Med Sci Sport. 1975;7:280–3.

Seiberl W, Power GA, Herzog W, Hahn D. The stretch-shortening cycle (SSC) revisited: residual force enhancement contributes to increased performance during fast SSCs of human m. adductor pollicis. Physiol Rep. 2015;3(5):e12401.

Strass D. Force-time and electromyographical characteristics of arm shoulder muscles in explosive type force production in sprint swimmers. J Swim Res. 1991;7:19–27.

Sugi H, Tsuchiya T. Stiffness changes during enhancement and deficit of isometric force by slow length changes in frog skeletal muscle fibres. J Physiol. 1988;407:215–29.

Toonstra J, Mattacola CG. Test-retest reliability and validity of isometric knee-flexion and -extension measurement using 3 methods of assessing muscle strength. J Sport Rehabil. 2013;22(1)

Van Driessche S, Van Roie E, Vanwanseele B, et al. Age-Related decline in leg-extensor power development in single- versus multijoint movements. Exp Gerontol. 2018a;110:98–104.

Van Driessche S, Delecluse C, Bautmans I, et al. Age-related differences in rate of power development exceed differences in peak power. Exp Gerontol. 2018b;101:95–100.

van Ingen Schenau GJ, Bobbert MF, De Haan A. Mechanics and energetics of the stretch-shortening cycle: a stimulating discussion. J Appl Biomech. 1997a;13:484–96.

van Ingen Schenau GJ, Bobbert MF, De Haan A. Does elastic energy enhance work and efficiency in the stretch-shortening cycle? J Appl Biomech. 1997b;13:389–415.

Verchoshansky YV. Componenti e struttura dell'impegno esplosivo di forza. SdS. 1996;34:15–21.

Viitasalo JT, Aura O. Seasonal fluctuation of force production in high jumpers. Can J Appl Sport Sci. 1984;9:209–13.

Viitasalo JT, Saukkonen S, Komi PV. Reproducibility of measurements of selected neuromuscular performance variables in man. Electromyogr Clin Neurophysiol. 1980;20:487–501.

Viitasalo JT, Hakkinen K, Komi PV. Isometric and dynamic force production and muscle fiber composition in man. J Hum Mov Stud. 1981;7:199–209.

Walshe AD, Wilson GJ, Ettema GJ. Stretch shorten cycle compared with isometric preload: contributions to enhanced muscular performance. J Appl Physiol. 1998;84:97–106.

Webber SC, Porter MM. Reliability of ankle isometric, isotonic, and isokinetic strength and power testing in older women. Phys Ther. 2010;90:1165–75.

Wilk KE, Keirns MA, Andrews JR, et al. Anterior cruciate ligament reconstruction rehabilitation: a six month follow up of isokinetic testing in recreational athletes. Isokinet Exerc Sci. 1991;1:1–2.

Wilson GJ, Murphy AJ. Quanta forza? Sport Med. 1997;5:21–9.

Wilson GJ, Wood GA, Elliot BC. Optimal stiffness of series elastic component in a stretch-shorten cycle activity. J Appl Physiol. 1991;70:825–33.

Wilson GJ, Murphy AJ, Pryor JF. Musclo-tendinous stiffness: its relationship to eccentric, isometric and concentric performance. J Appl Physiol. 1994;76:2714–9.

Wilson GJ, Lyttle D, Ostrowski KJ, Murphy AJ. Assessing dynamic performance: a comparison of rate of force development tests. J Strength Condit Assoc. 1995;9:176–81.

Young WB, Bilby GE. The effect of voluntary effort to influence speed of contraction on strength, muscular power and hypertrophy development. J Strength Cond Res. 1993;7:172–8.

7 The Use of Global Positioning System in the Return to Play Decision-Making Process

7.1 Introduction

The Global Positioning System (GPS) is a satellite-based navigational technology initially devised in the military environment. Since GPS technology enables three-dimensional movement of a subject or a group to be tracked over time in air, aquatic or land-based environments, its utility in the military context appears very clear (Aaltonen and Laarni 2017). The rapid development of GPS technology has made its use possible also in sports. It was first used in a sport activity in 1997 (Schutz and Chambaz 1997), and it is today widely used in several sport activities, such as soccer, Australian football, rugby, cricket, and hockey (Cummins et al. 2013). The GPS technology in team sports allows the measurement of players' position, velocity, and movement patterns. The measurement of a player's movements by means of the GPS allows to quantify, in the most objective manner possible, the subject's physiologic demand, the training intensity and the competition, and the training performance (McLellan et al. 2011). Basically, the GPS enables the measurement of a player's movement patterns, speed, and distance travelled, as well as the number of accelerations and decelerations. Furthermore, by adding a triaxial accelerometer, data concerning the physical workloads can be recorded (Cummins et al. 2013). The triaxial accelerometer allows the measurement of a composite vector magnitude that is expressed as a G-force. This is possible by recording the sum of accelerations measured in three axes, that is, the X, Y, and Z planes (Waldron et al. 2011). The data recorded by the GPS are used to calculate the player's movement pattern (the so-called external loads) and their physiological response to competition and training load (the so-called internal load). Beyond these data, both the number and the intensity of physical contacts and collisions can be calculated between athletes and objects, other athletes, and surfaces, through quantification of body load and impact measure. In particular the body load, which is quantified in G-force, represents the collection of all the forces imposed to the player (i.e., acceleration, deceleration, changes of direction, and impacts). GPS devices are currently manufactured with 1-, 5-,10-, and 15-Hz sampling rates (Johnston et al. 2014), where GPS devices having a higher-frequency rate provide greater reliability in the measurement (Jennings et al. 2010; Aughey 2011). However, some studies show no additional benefits when increasing to 15-Hz sampling rate (Johnston et al. 2014; Scott and Scott 2016). In any case, independently from the sampling rate of the device, the validity of the distance measured improves in relationship with the activities duration (Aughey 2011). For example, the standard of error is reduced in average by 67% when comparing sprinting over 40 and 10 m distances (Jennings et al. 2010). On the contrary, the reliability of GPS decreases with the increased

G. N. Bisciotti et al., *Return-to-Play after Lower Limb Muscle Injury in Football*,
https://doi.org/10.1007/978-3-030-84950-4_7

velocity of the movement (Jennings et al. 2010). Indeed, the GPS reliability is reduced when the movement speed is higher than 20 km $\cdot$ h^{-1} (Gray et al. 2010), probably because rapid changes in velocity happen when the football player moves at higher speeds (Jennings et al. 2010). In general, in literature the majority of the studies concluded that GPS devices show a sufficient level of both validity and retest reliability during the acquisition of movement patterns performed at lower speeds and over increased distance efforts. Such devices show less reliability during short-duration high-speed straight line running and changes of direction. Those situations do represent a limit for the accuracy in the acquisition of the abovementioned parameters in team sports. For this reason, caution is suggested in the interpretation of the sprints and rapid changes in both direction and velocity (Cummins et al. 2013). Moreover, further research on the validity and reliability in quantifying impacts in collision sports is needed.

7.2 Acquisition of Different Variables

The total distance covered, during training, is generally adopted as a measure of session or competition volume. The relative distance (i.e., the distance analyzed as a function of exposure time and expressed in m • min^{-1}) is usually considered as an indicator of the intensity of training sessions. Concerning the variables derived from speed data, the peak speed is generally considered to quantify the maximum effort, while the average speed is an indicator of external training load intensity. One of the most used "composite variable" is the "metabolic power," using a threshold of >25 W $\cdot$ kg^{-1} as index of high metabolic power load (Di Prampero et al. 2005, 2015; Osgnach et al. 2010). Other frequently used variables are, in order of importance, represented by Rago et al. (2019):

1. Speed-derived variables
2. Arbitrary speed zones
3. Average speed/distance for minute
4. Peek speed
5. Individualized speed zones
6. Speed intensity
7. Acceleration-derived variables
8. Arbitrary acceleration/deceleration zones
9. Player load
10. Impacts
11. Body load
12. Dynamic stress load
13. Metabolic power load
14. Repeated high-intensity efforts
15. High-intensity burst
16. Energy expenditure

7.3 The Classification of the Intensity Zones

Running is normally classified according to speed, and for male football players, the range from 20 to 25 km $\cdot$ h^{-1} is generally considered high-speed running. Below are the speed categories:

1. Walking (range ~0 to <~5.4 km/h)
2. Jogging (range ~5.5 to <~10.8 km/h)
3. Low speed running (range ~10.9 to <~14.4 km/h)
4. Intermediate speed running (range ~14.5 to <~19.8 km/h)
5. High speed running (range ~19.9 to <~25.2 km/h)
6. Maximum speed running (≥~25.2 km/h)

Acceleration and consequently deceleration are yet another measure of high-intensity exertions in team sport. In the same manner as for speed, for male football players, the different values of acceleration and deceleration are categorized in zones. The interval from 2 to 3 m • s^{-2} is usually considered a high, or intense, acceleration/deceleration. Yet, given the poor reliability of acceleration/deceleration values with increasing intensity thresholds, these data should be considered with great caution. Indeed, a problem not yet solved in the calculation of the acceleration and deceleration values is represented by the unknown speed from which accelerations/decelerations actually begin. However, the literature offers different studies focused on the GPS in soccer with a large number of speed, acceleration, and deceleration zones. For this reason, drawing objective conclusions concerning an agreement on common intensity zones is extremely difficult. This lack of consensus is particularly evident in the female football player population (Scott and Lovell 2018). Indeed, in such population today, the scarce literature available does not allow to establish any common speed zones (Rago et al. 2019). Probably, the best (and most practical) solution is to use arbitrary intensity zones based on the individual characteristics of each player. Indeed, the use of individualized intensity zones might furnish better information about the player's training workload adaptation rather than the use of commonly arbitrary intensity zones. In any case, it is true on one hand that the difference in field position and individual characteristics prevents the development of a classification with arbitrary zones, but on the other hand, the use of individualized intensity zones poses other important technical problems (Waldron et al. 2011). In addition, with the use of current GPS devices, limited evidence is available concerning vertical movements, changes of direction, and the number of jumps of a given height (Bloomfield et al. 2007; Rago et al. 2019).

7.4 GPS for Evaluating Impact and Collision Data

GPS devices are able to measure the forces upon the body using integrated accelerometer sensing devices (Kelly et al. 2012). The impact intensity is generally subdivided into six impact zones based upon the grading system provided by GPS manufacturers or in accordance with the literature (Cunniffe et al. 2009; Cummins et al. 2013). However, the accuracy of GPS technology concerning the evaluation of impact and collision data has not been fully elucidated and examined yet. In any case, with state-of-the-art GPS technology, the analysis of loads experienced through tackles or collisions is still relatively complex. Indeed, a reliable analysis of tackle and contact still requires manual pairing of GPS loads and video footage.

7.5 The Use of GPS in Different Age Groups and in Different Levels of Athletic Performance

Different ages of the athletes require the addition of appropriate speed zones so as to more accurately analyze the work rate patterns in relationship with the age. The speed zone classifications normally used to analyze the performance of adult athletes are not suitable for an objective and correct analysis of the prepubescent athletes. The reason lies in the obvious physiologic, biomechanical, and metabolic differences correlated to the age. One of the most important physiological differences between young and adult athletes is that during the prepubertal and pubertal age, the subjects exhibit lower energy reserve between submaximal and maximal aerobic exercise. This leads to an increased locomotion cost because of which, for a given running speed, the prepubertal and pubertal subjects work at a higher percentage of their maximal aerobic capacity in comparison to the adult subjects (Ebbeling et al. 1992). The important metabolic differences during exercise found in children, prepubescent, adolescents, and adults are attributable to five main physiological factors:

1. Children, prepubescent, and adolescents have, in comparison to the adults, shorter legs requiring a higher stride frequency and smaller stride length. This implies that they show a lower walking/running economy (Rowland et al. 1987).
2. Children, prepubescent, and adolescents have a less efficient running mechanics. Indeed, they show higher peak ground reaction forces, greater braking forces, and greater vertical movements (Grieve and Ruth 1966; Ebbeling et al. 1992).
3. The abovementioned age categories have a less efficient co-contraction of antagonist muscles, thus showing a suboptimal

neuromuscular control in the synchronization between the different muscle groups (Frost et al. 1997).
4. The subjects belonging to those age categories have a mass–speed imbalance which is particularly disadvantageous for the lighter children (Davies 1980; Thorstensson 1986).
5. The energy cost of the exercise steadily decline throughout childhood and into late adolescence (Ariens et al. 1997).

In addition to the problem related to different age groups, different levels of athletic ability constitute an issue too. Professional and nonprofessional athletes obviously have different athletic performances, which may also require the addition of other reduced speed zones, according to the athletes' different ability and skill level (Hartwig et al. 2011; Venter et al. 2011).

7.6 The Use of GPS for Injuries Prevention

The use of GPS to monitor the training workload may be interesting to determine the maximum training load that the athlete can sustain before a remarkable increase in the risk of injury takes place (Gabbett and Domrow 2007). Some authors show that in professional athletes 42% of illnesses and 40% of injuries could be explained by a previous excessive spike in the training load (Piggot et al. 2009). Indeed, the training workload should not exceed the individual's exercise tolerance and his/her capability of recovery. Recently in literature some interesting injury prevention models have appeared, based on the record of training workload data by means of GPS technology (Gabbett 2010; Windt and Gabbett 2017). Consequently, GPS technology can be used to record, check, and regulate the individual training workload. The fatigue effect is an important risk factor for injury, and the possibility to monitor fatigue development during a football match with GPS has been recently examined (Randers et al. 2010). The match-fatigue index is usually based on the difference in high-intensity running between the first and last 15 min of the match. Despite the fact that using the initial 15 min of the match as a benchmark for the determination of match-related fatigue is highly questionable (Lovell et al. 2013), the data recorded by the GPS during the match are not significantly different from either a video-based time-motion analysis system or a semiautomatic multiple-camera system (Cummins et al. 2013). However, beyond these interesting data, the use of GPS for injury prevention requires further research.

7.7 Conclusions

The introduction of GPS technology in football has allowed coaches, physical trainers, and physicians to be more objective in quantifying the external workload. The use of GPS parameters could help the medical department to determine the correct timing for return to play, giving important information concerning the physiological conditions of the player, helping to better characterize his training drills. However, the use of GPS technology provides many training load data, whose interpretation in football codes and in soccer is today very wide and rather inhomogeneous. Indeed, the collection of such an important amount of data, if not rationally interpreted and used, can generate a lot of confusion. In particular, the lack of uniformity in classifying speed, acceleration, and metabolic power thresholds limits the comparisons between the different studies present in literature today. Furthermore, the lack of consistent information in female football population should be underlined. Another important point is the necessity to investigate the physiological demands and characteristics of young football players, thus being able to modify the speed zone for a more accurate interpretation of the data reflecting the age and skill level of these athletes. Here, the GPS data may be very useful for coaches, physical trainers, and physicians in the definition of the most correct methods for the training of young football players.

References

Aaltonen I, Laarni J. Field evaluation of a wearable multimodal soldier navigation system. Appl Ergon. 2017;63:79–90.

Ariens GA, Mechelen W, Kemper HC, et al. The longitudinal development of running economy in males and females aged between 13 and 27 years: the Amsterdam Growth and Health Study. Eur J Appl Physiol Occup Physiol. 1997;76(3):214–20.

Aughey RJ. Applications of GPS technologies to field sports. Int J Sports Physiol Perform. 2011;6(3):295–310.

Bloomfield J, Polman R, O'Donoghue P. Physical demands of different positions in FA Premier League Soccer. J Sports Sci Med. 2007;6(1):63–70.

Cummins C, Orr R, O'Connor H, West C. Global positioning systems (GPS) and microtechnology sensors in team sports: a systematic review. Sports Med. 2013;43(10):1025–42.

Cunniffe B, Griffiths H, Proctor W, Jones KP, Baker JS, Davies B. Illness monitoring in team sports using a Webbased training diary. Clin J Sport Med. 2009;19(6):476–81.

Davies CT. Metabolic cost of exercise and physical performance in children with some observations on external loading. Eur J Appl Physiol Occup Physiol. 1980;45(2–3):95–102.

Di Prampero PE, Fusi S, Sepulcri L, Morin JB, Belli A, Antonutto G. Sprint running a new energetic approach. J Exp Biol. 2005;208(Pt 14):2809–16.

Di Prampero PE, Botter A, Osgnach C. The energy cost of sprint running and the role of metabolic power in setting top performances. Eur J Appl Physiol. 2015;115(3):451–69.

Ebbeling C, Hamill J, Freedson P, et al. An examination of efficiency during walking in children and adults. Pediatr Exerc Sci. 1992;4(1):36–49.

Frost G, Dowling J, Dyson K, et al. Cocontraction in three age groups of children during treadmill locomotion. J Electromyogr Kinesiol. 1997;7(3):179–86.

Gabbett TJ. The development and application of an injury prediction model for noncontact, soft-tissue injuries in elite collision sport athletes. J Strength Cond Res. 2010;24(10):2593–603.

Gabbett TJ, Domrow N. Relationships between training load, injury, and fitness in sub-elite collision sport athletes. J Sports Sci. 2007;25(13):1507–19.

Gray AJ, Jenkins D, Andrews MH, et al. Validity and reliability of GPS for measuring distance travelled in field-based team sports. J Sports Sci. 2010;28(12):1319–25.

Grieve DW, Ruth GJ. The relationships between length of stride, step frequency, time of swing and speed of walking for children and adults. Ergonomics. 1966;9(5):379–99.

Hartwig T, Naughton G, Searl J. Motion analyses of adolescent Rugby Union players: a comparison of training and game demands. J Strength Cond Res. 2011;25(4):966–72.

Jennings D, Cormack S, Coutts AJ, et al. The validity and reliability of GPS units for measuring distance in team sport specific running patterns. Int J Sports Physiol Perform. 2010;5(3):328–41.

Johnston RJ, Watsford ML, Kelly SJ, Pine MJ, Spurrs RW. Validity and interunit reliability of 10 Hz and 15 Hz GPS units for assessing athlete movement demands. J Strength Cond Res. 2014;28(6):1649–55.

Kelly D, Coughlan FG, Green SB, et al. Automatic detection of collisions in elite level Rugby union using a wearable sensing device. Sports Eng. 2012;15:81–92.

Lovell R, Barrett S, Portas M, et al. Re-examination of the post half-time reduction in soccer work-rate. J Sci Med Sport. 2013;16:250–4.

McLellan CP, Lovell DI, Cass GC. Performance analysis of elite Rugby League match play using global positioning systems. J Strength Cond Res. 2011;25(6):1703–10.

Osgnach C, Poser S, Bernardini R, Rinaldo R, di Prampero PE. Energy cost and metabolic power in elite soccer: a new match analysis approach. Med Sci Sports Exerc. 2010;42(1):170–8.

Piggot B, Newton M, McGuian M. The relationship between training load and incidence of injury and illness over a pre season at an Australian Football League club. J Aus Strength Cond. 2009;17(3):4–17.

Rago V, Brito J, Figueiredo P, Costa J, Barreira D, Krustrup P, Rebelo A. Methods to collect and interpret external training load using microtechnology incorporating GPS in professional football: a systematic review. Res Sports Med. 2019;22:1–22.

Randers M, Mujikab I, Hewitt A, et al. Application of four different football match analysis systems: a comparative study. J Sports Sci. 2010;28(2):171–82.

Rowland TW, Auchinachie JA, Keenan TJ, et al. Physiologic responses to treadmill running in adult and prepubertal males. Int J Sports Med. 1987;8(4):292–7.

Schutz Y, Chambaz A. Could a satellite-based navigation system(GPS) be used to assess the physical activity of individuals on earth? Eur J Clin Nutr. 1997;51(5):338–9.

Scott D, Lovell R. Individualisation of speed thresholds does not enhance the dose-response determination in football training. J Sports Sci. 2018;36(13):1523–32.

Scott MT, Scott TJ, Kelly VG. The validity and reliability of global positioning systems in team sport: a brief review. J Strength Cond Res. 2016;30(5):1470–90.

Thorstensson A. Effects of moderate external loading on the aerobic demand of submaximal running in men and 10 year-old boys. Eur J Appl Physiol Occup Physiol. 1986;55(6):569–74.

Venter R, Opperman E, Opperman S. The use of global positioning system (GPS) tracking devices to access movement demands and impacts in under-19 Rugby Union match play. Afr J Phys Health Ed Rec Dance. 2011;17(1):1–8.

Waldron MT, Highton C, Worsolf J, et al. Movement and physiological match demands of elite Rugby League using portable global positioning systems. J Sports Sci. 2011;29(11):1223–30.

Windt J, Gabbett TJ. How do training and competition workloads relate to injury? The workload-injury aetiology model. Br J Sports Med. 2017;51(5):428–35.

8 Return to Training and Return to Play Following Quadriceps Injury

8.1 Anatomical Description

The quadriceps (Fig. 8.1) is the largest muscle in the anterior region of the thigh and is composed of four heads:

1. Rectus femoris (RF)
2. Vastus lateralis (VL)
3. Vastus medialis (VM)
4. Vastus intermedius (VI)

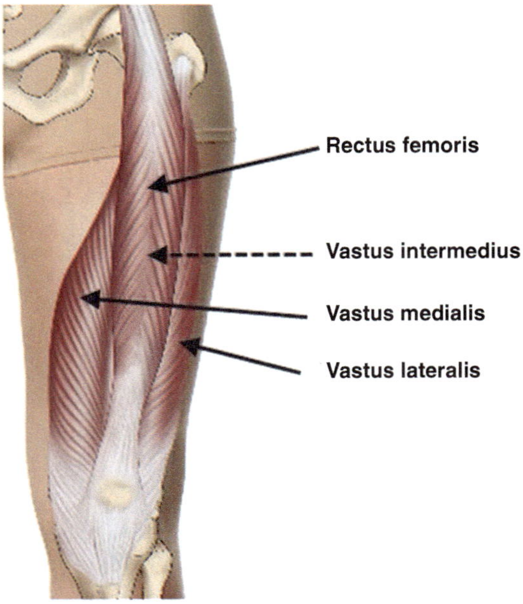

Fig. 8.1 Anterior schematic view of the quadriceps femoris

Three of these muscles (vastus lateralis, medialis, and intermedius) are monoarticular, and the only biarticular muscle of the quadriceps complex is the rectus femoris. The RF is a long, fleshy muscle located in the anterior compartment of the thigh. It is fusiform in shape with superficial fibers that are bipenniform and deep fibers that run straight to the deep aponeurosis (Bordoni and Varacallo 2018). RF is the most superficial of the quadriceps muscles, and it is superficially crossed by the sartorius muscle. The RF originates with three tendons: the direct, the indirect or reflected, and the recurrent. The direct tendon arises from the anterior aspect of the inferior iliac spine, the indirect tendon originates from the upper acetabular crest and the posterolateral capsule, and the recurrent tendon originates from the capsule anterior aspect. The two direct and indirect tendons form the conjoint tendon 2 cm distal to their origin. However, each tendon retains its own identity. The direct tendon remains superficial on the ventral muscle surface, while the indirect tendon extends throughout the distal two-thirds of the muscle belly and lies superficially on the dorsal surface of the muscle (Chammout and Skinner 1986). Furthermore, some studies based on cadaveric dissections showed that the proximal tendon has two components: one superficial and one deep, rather than being a single superficial tendon (Hasselman et al. 1995). The RF shows a more complex muscle architecture than a simple bipennate structure as commonly described.

G. N. Bisciotti et al., *Return-to-Play after Lower Limb Muscle Injury in Football*,
https://doi.org/10.1007/978-3-030-84950-4_8

Indeed, in its proximal third muscle-tendon unit, its muscle fibers arise from the superficial direct tendon head and then travel distally and directly to the posterior distal tendon. For this reason, the proximal third of RF shows a unipennate structure. On the contrary, distal to the superficial proximal direct tendon, the muscle fibers arise from indirect tendon head and travel distally in the posterior direction to the posterior tendon, creating in this manner a bipennate structure (Hasselman et al. 1995). This complex muscle architecture creates a so-called "muscle in the muscle" aspect in the RF distal half (Hasselman et al. 1995). Some authors consider this "muscle in the muscle structure" a true different smaller muscle within the RF (Hasselman et al. 1995). Finally, the two heads (direct and indirect) merge, as previously said, to become the conjoined tendon below their origin (Kumaravel et al. 2018).

The RF merging with the VL, VM, and VI tendons inserts distally on the tibial tuberosity. It is innervated by the femoral nerve (L2-L4), and its vascular supply comes from the descending branch of the lateral femoral circumflex artery. Its main action is to extend the knee, while it also participates, with the sartorius and iliopsoas, in the flexion of the thigh.

The VL is the largest component of the quadriceps complex. It originates from the femur via a broad aponeurosis inserted on the upper part of the intertrochanteric line, the anterior and inferior margins of the great trochanter, the gluteal tuberosity, the lateral portion of the *linea aspera*, and lateral intermuscular septum. The VL distal tendon is inserted on the superolateral margin of the patella and on the lateral condyle of the tibia. It is innervated by two branches of the femoral nerve (L3-L4), and its vascular supply comes from the descending branch of the lateral femoral circumflex artery. VL inserts with an aponeurotic tendon on the RF tendon, on the superolateral margin of the patella and on the lateral condyle of the tibia. With its action it extends the leg.

The VM originates from the medial aspect of the *linea aspera* and inserts distally with an aponeurotic tendon on the tendon of the RF, on the superior-medial margin of the patella and on the medial condyle of the tibia. With its action, it extends the leg and stabilizes the patella opposing its tendency to dislocate outward. VM is the most important dynamic stabilizer of the knee joint. It is innervated by two branches of the femoral nerve (L3, L4), and its vascular supply comes from the descending branches of the lateral circumflex artery of the thigh.

VI originates from the femoral diaphysis, from the lower half the lateral aspect of the *linea aspera* and from the intermuscular lateral septum. Its fibers end in a superficial aponeurosis which joins the deep surface of the RF, VL, and VM tendons. With its action, it extends the leg. VI is innervated on the surface by two branches of the femoral nerve (L3, L4). Its vascular supply comes from the descending branches of the lateral circumflex artery of the thigh. In 2016, a group of Swiss researchers (Grob et al. 2016) discovered a new quadriceps head that was called the "Tensor of the Vastus Intermedius" (TVI). This "new" muscle is located between the VI and the VL. It originates in the proximal anterior part of the femur and continues with an aponeurosis that merges with the quadriceps tendon, therefore being a very short muscle. It is assumed that TVI may have an important function in the control of patella, or in the synergism with the VI action creating additional tension, hence being called the tensor of the vast intermediate. In the considered anatomical specimen, 42% had a TVI that did not join with other tendons, 23% of the TVI tendons shared an aponeurosis with the VI, 19% with the VL, and 15% with both VI and VL.

8.2 Epidemiological Notes

In football, the majority (~88%) of quadriceps femoris injuries assessed with MRI examination involve the RF (Hägglund et al. 2013a; Hägglund et al. 2013b). The risk of suffering from this type of injury is higher during competition than during training (1.1 versus 0.3 per 1000 h of exposure) (Bengtsson et al. 2013; Hägglund et al. 2013a; Hägglund et al. 2013b); 62% of rectus femoris lesions are recorded during the first half of the match, and the peak risk is observed between the 16th and 45th mins of play

(Bengtsson et al. 2013; Hägglund et al. 2013a). The most common mechanism of injury is during the kicking motion (~28% of injuries). The rate of reinjury is approximately 13% (Deehan et al. 2007; Fousekis et al. 2011; Mueller-Wohlfahrt et al. 2013), and a team of 25 players should expect on average three lesions of the rectus femoris per season, resulting in a total time loss of around 50 days (Hägglund et al. 2013a, b).

8.3 Clinical and Imaging Assessments for RTT

In literature, there are few studies that refer to decision-making process (DMP) for RTT and RTP following indirect injury of QF. Indeed, almost all of the studies focus on hypotonotrophy and/or iatrogenic damage to the QF following arthroscopic ACL reconstruction.

The tests suggested by the CC for QF indirect injury are the following.

8.3.1 General Assessment

Below are the Consensus Conference recommendations for clinical and imaging assessments for RTT following quadriceps muscle injury:

1. Absence of clinical symptoms (Kvist 2004; Malliaropoulos et al. 2011; Delvaux et al. 2014).
2. Absence of pain or tenderness during muscle palpation (Kvist 2004; Zambaldi et al. 2017; Bisciotti and Volpi 2018; Reurink et al. 2014).
3. Absence of pain on passive and active stretching (Witvrouw et al. 2003; Bisciotti and Volpi 2018).
4. Absence of pain on isometric, concentric and eccentric contraction (Bisciotti and Volpi 2018).
5. Completion of the prescribed rehabilitation program (Reurink et al. 2014).
6. MRI and US imaging assessment points will be specified in Chap. 5: “The Role of Imaging in the Return to Training and Return to Play Decision-Making Process.”
7. Subjective feelings of the player taken into account (i.e., assess levels of anxiety, apprehension, fear of failure, and/or fear of reinjury) (McCarty et al. 2004; Bauman 2005; Glazer 2009; Langford et al. 2009; Clover and Wall 2010; Delvaux et al. 2014).

8.3.2 Specific Assessment

Passive quadriceps stretch test (Witvrouw et al. 2003; Bouvier et al. 2017).

8.4 Laboratory Tests for RTT

After quadriceps injury, the laboratory tests for RTT recommended by the CC are:

1. Quadriceps muscles strength assessment by means of dynamometric tests. The values of dynamometric tests must be >90% of prelesional or contralateral values (Bisciotti 2015).
2. Synchro plates test (Bisciotti et al. 2016).

8.5 Field Tests for RTT

The following are the field tests recommended by the CC to determine readiness to RTT after quadriceps injury:

1. Illinois Agility Test (Hachana et al. 2013; Raya et al. 2013; Bisciotti 2015; Negra et al. 2017; Bisciotti and Volpi 2018).
2. Braking test (Bisciotti 2015; Bisciotti and Volpi 2018).
3. Kicking test (Bisciotti 2015; Bisciotti and Volpi 2018).

No previous validation studies were identified on the use of field tests to inform RTT and RTP following quadriceps injuries. However, the CC considered an RTP test checklist for athletes who suffered a lower extremity injury, set out in a 2013 Delphi study (Haines et al. 2013). The tests were recommended based on expert opinions

(GRADE evidence level V). Furthermore, please remember that the Illinois Agility Test is asymmetric (Hachana et al. 2013; Raya et al. 2013; Bisciotti 2015; Negra et al. 2017; Bisciotti and Volpi 2018), Therefore the Consensus Conference recommends its execution in the modified format formulated by Rouissi et al. (2016).

8.6 RTP Tests after QS Injury

The criteria for establishing RTP-DMP are based on performance evaluation, which must chronologically follow the RTT-DMP. Therefore, once the athlete has obtained a positive judgment for RTT, the period of training registration with GPS technology, which is the judgment context for RTP, can begin.

In this regard, the CC recommend the following points:

1. The data acquisition period must start from the first day of RTP and last 7–10 days.
2. During this period, the performance data must be systematically recorded via the GPS system.
3. It is necessary to identify some "typical" sessions, substantially overlapping each other, which can be deduced from the pre-lesional period.

As already mentioned in Chap. 4, the three evaluation categories are:

1. Quantitative evaluation.
2. Qualitative evaluation.
3. Parameter analysis.

All the indices described in Chap. 4 as related to the abovementioned categories must be analyzed. The reference value below which the positive judgment for RTP is postponed is set at a maximum difference of 10% between the pre-injury data and the data recorded during the acquisition period following RTT. Concerning the control of aerobic fitness, the CC experts suggest that the RTP-DMP be implemented by means of a test for VO_2max determination (Heck et al. 1985; Chamari et al. 2004; Chamari et al. 2005; Badawy and Muaidi 2018), in which the player must record a value of VO_2max, and consequently of VAM, equal to at least 90% of the pre-injury values.

References

Badawy MM, Muaidi QI. Aerobic profile during high-intensity performance in professional Saudi Athletes. Pak J Biol Sci. 2018;21(1):24–8.

Bauman J. Returning to play: the mind does matter. Clin J Sport Med. 2005;15:432–5.

Bengtsson H, Ekstrand J, Hagglund M. Muscle injury rates in professional football increase with fixture congestion: an 11-year follow-up of the UEFA champions league injury study. Br J Sports Med. 2013;47: 743–7.

Bisciotti GN. Return to play after a muscle lesion. In: Volpi P, editor. Arthroscopy in sport. Cham: Springer Edition; 2015.

Bisciotti GN, Quaglia A, Belli A, et al. Return to sports after ACL reconstruction: a new functional test protocol. Muscles Ligaments Tendons J. 2016;06:499–509.

Bisciotti GN, Volpi P. Return to play. In: Volpi P, editor. Football doctor manual. Trento: Edra Edition; 2018. p. 247–59.

Bordoni B, Varacallo M. Anatomy, bony pelvis and lower limb, thigh quadriceps muscle. Treasure Island: StatPearls Publishing; Dec 15, 2018.

Bouvier T, Opplert J, Cometti C, et al. Acute effects of static stretching on muscle-tendon mechanics of quadriceps and plantar flexor muscles. Eur J Appl Physiol. 2017;117:1309–15.

Chamari K, Hachana Y, Ahmed YB, Galy O, Sghaïer F, Chatard JC, Hue O, Wisløff U. Field and laboratory testing in young elite soccer players. Br J Sports Med. 2004 Apr;38(2):191–6.

Chamari K, Moussa-Charai I, Boussaïdi L, Hachana Y, Kauech F, Wisløff U. Appropriate interpretation of aerobic capacity: allometric scaling in adult and young soccer players. Br J Sports Med. 2005 Feb;39(2):97–101.

Chammout MO, Skinner HB. The clinical anatomy of commonly injured muscle bellies. J Trauma. 1986;26:549–52.

Clover J, Wall J. Return-to-play criteria following sports injury. Clin Sports Med. 2010;29:169–75.

Deehan DJ, Bell K, McCaskie AW. Adolescent musculoskeletal injuries in a football Academy. J Bone Joint Surg Br. 2007;89- B:5–8.

Delvaux F, Rochcongar P, Bruyère O, Bourlet G, Daniel C, Diverse P, Reginster JY, Croisier JL. Return-to-play criteria after hamstring injury: actual medicine practice in professional soccer teams. J Sports Sci Med. 2014 Sep 1;13(3):721–3.

Fousekis K, Tsepis E, Poulmedis P, et al. Intrinsic risk factors of non-contact quadriceps and hamstring strains in soccer: a prospective study of 100 professional players. Br J Sports Med. 2011;45:709–14.

Glazer DD. Development and preliminary validation of the injury- psychological readiness to return to sport (I-PRRS) scale. J Athl Train. 2009;44:185–9.

Grob K, Ackland T, Kuster MS, Manestar M, Filgueira L. A newly discovered muscle: the tensor of the vastus intermedius. Clin Anat. 2016 Mar;29(2): 256–63.

Hachana Y, Chaabene H, Nabli MA, et al. Test-retest reliability, criterion-related validity, and minimal detectable change of the Illinois agility test in male team sport athletes. J Strength Cond Res. 2013;27: 2752–9.

Hägglund M, Walden M, Ekstrand J. Risk factors for lower extremity muscle injury in professional soccer: the UEFA injury study. Am J Sports Med. 2013b;41:327–35.

Hägglund M, Walden M, Magnusson H, et al. Injuries affect team performance negatively in professional football: an 11-year followup of the UEFA champions league injury study. Br J Sports Med. 2013a;47:738–42.

Haines S, Baker T, Donaldson M. Development of a physical performance assessment checklist for athletes who sustained a lower extremity injury in preparation for return to sport: a Delphi study. Int J Sports Phys Ther. 2013;8:44–53.

Hasselman CT, Best TM, Hughes C, Martinez S, Garrett WE Jr. An explanation for various rectus femoris strain injuries using previously undescribed muscle architecture. Am J Sports Med. 1995;23(4):493–9.

Heck H, Mader A, Hess G, Mucke S, Muller R, Hollmann W. Justification of the 4-mmol/l lactate threshold. Int J Sports Med. 1985;6:117–30.

Kumaravel M, Bawa P, Murai N. Magnetic resonance imaging of muscle injury in elite American football players: predictors for return to play and performance. Eur J Radiol. 2018 Nov;108:155–64.

Kvist J. Rehabilitation following anterior cruciate ligament injury: current recommendations for sport participation. Sports Med. 2004;34:296–80.

Langford JL, Webster KE, Feller JA. A prospective longitudinal study to assess psychological changes following anterior cruciate ligament reconstruction surgery. Br J Sports Med. 2009;43:377–8.

Malliaropoulos N, Isinkaye T, Tsitas K, et al. Reinjury after acute posterior thigh muscle injuries in elite track and field athletes. Am J Sports Med. 2011;39(2):304–10.

McCarty EC, Ritchie P, Gill HS, et al. Shoulder instability: return to play. Clin Sports Med. 2004;23:335–51.

Mueller-Wohlfahrt HW, Ueblacker P, Haensel L, et al. Muscle injuries in sports. New York: Thieme Editions; 2013. p. 42–3.

Negra Y, Chaabene H, Hammami M, et al. Agility in young athletes: is it a different ability from speed and power? J Strength Cond Res. 2017;31:727–35.

Raya MA, Gailey RS, Gaunaurd IA, et al. Comparison of three agility tests with male servicemembers: Edgren side step test, t-test, and Illinois agility test. J Rehabil Res Dev. 2013;50:951–60.

Reurink G, Goudswaard GJ, Tol JL, et al. MRI observations at return to play of clinically recovered hamstring injuries. Br J Sports Med. 2014;48(18):1370–6.

Rouissi M, Chtara M, Berriri A, et al. Asymmetry of the modified Illinois change of direction test impacts young elite soccer players' performance. Asian J Sports Med. 2016;7:e33598.

Witvrouw E, Danneels L, Asselman P, et al. Muscle flexibility as a risk factor for developing muscle injuries in male professional soccer players. Am J Sports Med. 2003;31:41–6.

Zambaldi M, Beasley I, Rushton A. Return to play criteria after hamstring muscle injury in professional football: a Delphi consensus study. Br J Sports Med. 2017;51:1221–6.

9 Return to Training and Return to Play Following Hamstring Injury

9.1 Anatomical Description

The hamstring muscle complex (Fig. 9.1) is comprised of three different muscles: semitendinosus, semimembranosus, and biceps femoris.

The biceps femoris (BF) is formed by a long and a short head. The BF long head originates from the ischial tuberosity, and its insertion is on the fibular head and lateral condyle of the tibia. Interestingly, there are reports regarding anatomical variants revealing cases where the semitendinosus and the long head of the biceps femoris appear from distinct tendinous origins (Koulouris and Connell 2005). Its function is the knee flexion, the lateral rotation of the tibia, and the hip extension. BF long head is innervated by the tibial nerve, and its vascular supply comes from the perforating branches of the deep femoral artery. The BF short head originates from the lateral lip of the *linea aspera*, and its insertion is on the fibular head and lateral condyle of the tibia. BF short head function is knee flexion and lateral rotation of the tibia. It is innervated by the fibular (common peroneal) nerve, and its vascular supply comes from the perforating branches of the deep femoral artery.

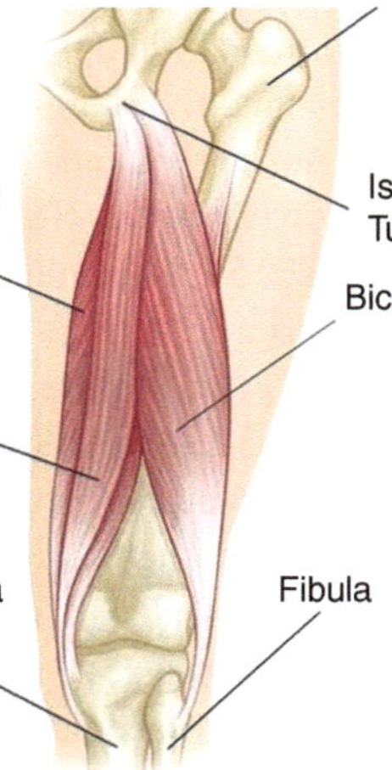

Fig. 9.1 Schematic representation of hamstring muscles

The semitendinosus muscle (ST) originates with a common tendon with the BF long head from ischial tuberosity and inserts on medial tibia (*pes anserinus*). Its function is the knee flexion, the hip extension, and the medial rotation of the tibia (with flexed knee). ST is innervated by the tibial nerve, and its vascular supply comes from the perforating branches of the deep femoral artery.

The semimembranosus muscle (SM) originates from the ischial tuberosity and inserts on medial tibial condyle. Its function is the knee flexion, the hip extension, and medial rotation of the tibia (with flexed knee). SM is innervated by the tibial nerve, and its vascular supply comes from the perforating branches of the deep femoral artery.

The hamstring (HS) complex, aside from playing an essential role in hip extension and knee flexion, is strongly involved in the gait cycle, during which the HS activation begins at the last

G. N. Bisciotti et al., *Return-to-Play after Lower Limb Muscle Injury in Football*,
https://doi.org/10.1007/978-3-030-84950-4_9

25% of the swing phase, generating extension force at the hip and resisting knee extension. The hamstring muscles also play an essential role as a dynamic stabilizer of the knee joint operating in synergy with the anterior cruciate ligament (ACL) to counteract the anterior translation of the tibia during the heel strike phase of the gait cycle (Koulouris and Connell 2005). Something interesting to note in the context of the HS functions is the so-called "Lombard's paradox" cycle (Lombard and Abbott 1907). Lombard's paradox is a paradoxical muscular contraction which happens when, rising to stand, both the hamstrings and quadriceps contract at the same time, even though they are antagonists to each other. In this paradoxical situation, the rectus femoris (the only biarticular muscle of the quadriceps muscle group) acting over the hip has a smaller hip moment arm than the HS. However, at the same time, the rectus femoris moment arm is greater over the knee than the HS knee moment. This situation means that contraction from both rectus femoris and HS results in a simultaneous hip and knee extension. Furthermore, the hip extension adds a passive stretch component to rectus femoris. However, this complex and paradoxically biomechanical situation is very useful allowing an efficient movement during the gait cycle (Lombard and Abbott 1907; Gregor et al. 1985; Andrews 1987).

The HS (specifically the ST, which is harvested together with the gracilis muscle) can be used as an autograft in ACL reconstruction. About this topic, the literature provides some studies demonstrating that HS autograft has an increased risk of knee laxity and functional hamstring weakness (Heiderscheit et al. 2010). Indeed, the surgical harvesting of the ST tendon is per se an iatrogenic damage to the HS muscle-tendon complex. Yet, literature also recognizes the regeneration of the harvested tendon (Ferretti et al. 2002), albeit that the distal insertion becomes in most cases the gastrocnemius fascia rather than the tibia (Cross et al. 1992; Simonian et al. 1997). Nevertheless, an ineffective scar regeneration may occur, which in athletes may result in a *locus minoris resistentiae* and therefore in a risk factor for reinjuries (Linklater et al. 2010).

9.2 Epidemiological Notes

Both the anatomical and functional nature of the HS predispose this muscular complex to injury, including the fact that the HS crosses two joints (knee and hip) and undergoes eccentric contraction during the gait cycle (Linklater et al. 2010). Indeed, for their own nature, they generate an important "spring effect" during the sport movements, on one hand enhancing the athletic performance and on the other hand increasing the injury risk (Linklater et al. 2010).

Furthermore, HS contains a relatively important percentage of fast-twitch fibers which represent another important risk factor (Garrett Jr et al. 1984; Volpi and Bisciotti 2019). HS injuries are the most frequent injury in football and account for about 17% of all football injuries (Hawkins et al., 2001; Morgan and Oberlander, 2001). The severity of these lesions ranges from simple delayed onset muscle soreness (DOMS) to complete avulsion (Kujala e coll., 1997). In football, a player risks 2.5 times more a HS indirect injury than a femoral quadriceps muscle indirect injury. A professional football team incurs an average of ten hamstring injuries per season (Hägglund et al. 2013a, b). This results in an average of 90 days of time lost to injury and on average between 15 and 21 matches lost per team per season. The incidence of HS injuries ranges from 0.87 to 0.96 per 1000 h of exposure (training and match) (Hägglund et al. 2013a, b). BF is the most frequently injured of the HS complex, followed by the SM and then by the ST (Brukner 2015). The majority of HS injuries receive nonsurgical treatment; however, HS tendon avulsion often requires surgical management (Folsom and Larson 2008). Ischial apophyseal avulsion fractures are extremely rare. Studies report that they account for between only 1.4 and 4% of all HS injuries (Liu et al. 2018). Ischial apophyseal avulsion fractures with over 1-cm displacement receive surgical treatment (Gidwani et al. 2004; Liu et al. 2018). In such case, early intervention is advised to decrease the risk of ischiofemoral impingement (Goldblatt et al. 2005). It is important to note that the secondary center of ossification of the ischium appears during puberty and is present until the 15th to 25th year, representing a weak point for the HS muscle-tendon unit (Linklater et al. 2010).

9.3 Clinical and Imaging Assessments for RTT

9.3.1 General Assessment

Below are the Consensus Conference recommendations for clinical and imaging assessments for RTT following hamstring injury:

1. Absence of clinical symptoms (Kvist 2004; Malliaropoulos et al. 2011; Delvaux et al. 2014).
2. Absence of pain or tenderness during muscle palpation (Kvist 2004; Zambaldi et al. 2017; Bisciotti et al. 2018; Reurink et al. 2014).
3. Absence of pain on passive and active stretching (Witvrouw et al. 2003; Bisciotti et al. 2018).
4. Absence of pain on isometric, concentric, and eccentric contraction (Bisciotti et al. 2018).
5. Completion of the prescribed rehabilitation program (Reurink et al. 2014).
6. MRI and US imaging assessment points will be specified in Chap. 5: "The Role of Imaging in the Return to Training and Return to Play Decision-Making Process."
7. Subjective feelings of the player taken into account (i.e., assess levels of anxiety, apprehension, fear of failure, and/or fear of reinjury) (McCarty et al. 2004; Bauman 2005; Glazer 2009; Langford et al. 2009; Clover and Wall 2010; Delvaux et al. 2014).

9.3.2 Specific Assessment

1. Passive straight leg raise test (Witvrouw et al. 2003; Delvaux et al. 2014; Kellis et al. 2015).
2. Dynamic flexibility H test (Askling et al. 2010).

9.4 Laboratory Tests for RTT

The following are the laboratory tests recommended prior to RTT:

1. Evaluation of hamstring muscle strength by dynamometric tests (isometric, isotonic, and isokinetic tests) (Croisier et al. 2002; Sanfilippo et al. 2013; Delvaux et al. 2014; Bisciotti et al. 2016).

9.5 Field Tests for RTT

The following are the field tests recommended to determine readiness to RTT after hamstring injury:

1. Illinois Agility Test (Hachana et al. 2013; Raya et al. 2013; Bisciotti 2015; Negra et al. 2017; Bisciotti et al. 2018).
2. Braking test (Bisciotti 2015; Bisciotti et al. 2018).
3. Backward running (Brumitt et al. 2013).

As in the case of quadriceps injuries, no previous validation studies were identified on the use of field tests to inform RTT and RTP. However, we considered an RTP test checklist for athletes who suffered a lower extremity injury, set out in a 2013 Delphi study (Haines et al. 2013). The tests were recommended based on expert opinions (GRADE evidence level V). Furthermore, the Illinois Agility Test is asymmetric (Hachana et al. 2013; Raya et al. 2013; Bisciotti 2015; Negra et al. 2017; Bisciotti and Volpi 2018); therefore the Consensus Conference recommends its execution in the modified format formulated by Rouissi et al. (2016).

9.6 RTP Tests after HS Injury

The criteria for establishing RTP-DMP are based on performance evaluation, which must chronologically follow the RTT-DMP. In other words, once the athlete has obtained a positive judgment for RTT, the period of registration of the tests, which are the judgment context for RTP, can begin.

In this regard, we recommend the following points:

1. The data acquisition period must start from the first day of RTP and last 7–10 days (depending upon the severity of the injury).
2. During this period, the performance data must be systematically recorded via the GPS system.
3. It is necessary to identify some "typical" sessions, substantially overlapping each other, which can be deduced from the pre-lesional period.

As already mentioned in Chap. 4, the three evaluation categories are:

1. Quantitative evaluation.
2. Qualitative evaluation.
3. Parameter analysis.

All the indices described in Chap. 4 as related to the abovementioned categories must be analyzed The reference value below which the positive judgment for RTP is postponed is set at a maximum difference of 10% between the pre-lesional data and the data recorded during the acquisition period following RTT. Moreover, regarding the control of aerobic fitness, the CC experts suggest that the RTP-DMP be implemented by means of a test for VO_2max determination (Heck et al. 1985; Chamari et al. 2004; Chamari et al. 2005; Badawy and Muaidi 2018), in which the player must record a value of VO_2max, and consequently of VAM, equal to at least 90% of the pre-injury values.

References

Andrews JG. The functional roles of the hamstrings and quadriceps during cycling: Lombard's paradox revisited. J Biomech. 1987;20(6):565–75.

Askling CM, Nilsson J, Thorstensson A. A new hamstring test to complement the common clinical examination before return to sport after injury. Knee Surg Sports Traumatol Arthrosc. 2010;18:1798–803.

Badawy MM, Muaidi QI. Aerobic Profile During High-intensity Performance in Professional Saudi Athletes. Pak J Biol Sci. 2018;21(1):24–8.

Bauman J. Returning to play: the mind does matter. Clin J Sport Med. 2005;15:432–5.

Bisciotti GN. Return to play after a muscle lesion. In: Volpi P, editor. Arthroscopy in sport. Berlin: Springer Edition; 2015.

Bisciotti GN, Quaglia A, Belli A, et al. Return to sports after ACL reconstruction: a new functional test protocol. Muscles Ligaments Tendons J. 2016;06:499–509.

Bisciotti GN, Volpi P. Return to play. In: Volpi P, editor. Football doctor manual. Trento: Edra Edition; 2018. p. 247–59.

Bisciotti GN, Volpi P, Amato M, Alberti G, Allegra F, Aprato A, Artina M, Auci A, Bait C, Bastieri GM, Balzarini L, Belli A, Bellini G, Bettinsoli P, Bisciotti A, Bisciotti A, Bona S, Brambilla L, Bresciani M, Buffoli M, Calanna F, Canata GL, Cardinali D, Carimati G, Cassaghi G, Cautero E, Cena E, Corradini B, Corsini A, D'Agostino C, De Donato M, Delle Rose G, Di Marzo F, Di Pietto F, Enrica D, Eirale C, Febbrari L, Ferrua P, Foglia A, Galbiati A, Gheza A, Giammattei C, Masia F, Melegati G, Moretti B, Moretti L, Niccolai R, Orgiani A, Orizio C, Pantalone A, Parra F, Patroni P, Pereira Ruiz MT, Perri M, Petrillo S, Pulici L, Quaglia A, Ricciotti L, Rosa F, Sasso N, Sprenger C, Tarantola C, Tenconi FG, Tosi F, Trainini M, Tucciarone A, Yekdah A, Vuckovic Z, Zini R, Chamari K. Italian consensus conference on guidelines for conservative treatment on lower limb muscle injuries in athlete. BMJ Open Sport Exerc Med. 2018 May 24;4(1):e000323. https://doi.org/10.1136/bmjsem-2017-000323.

Brukner P. Hamstring injuries: prevention and treatment-an update. Br J Sports Med. 2015 Oct;49(19):1241–4.

Brumitt J, Heiderscheit BC, Manske RC, et al. Lower extremity functional tests and risk of injury in division III collegiate athletes. Int J Sports Phys Ther. 2013;8:216–27.

Chamari K, Hachana Y, Ahmed YB, Galy O, Sghaïer F, Chatard JC, Hue O, Wisløff U. Field and laboratory testing in young elite soccer players. Br J Sports Med. 2004 Apr;38(2):191–6.

Chamari K, Moussa-Charai I, Boussaïdi L, Hachana Y, Kauech F, Wisløff U. Appropriate interpretation of aerobic capacity: allometric scaling in adult and young soccer players. Br J Sports Med. 2005 Feb;39(2):97–101.

Clover J, Wall J. Return-to-play criteria following sports injury. Clin Sports Med. 2010;29:169–75.

Croisier J-L, Forthomme B, Namurois M-H, et al. Hamstring muscle strain recurrence and strength performance disorders. Am J Sports Med. 2002;30:199–203.

Cross MJ, Roger G, Kujawa P, Anderson IF. Regeneration of the semitendinosus and gracilis tendons following their transection for repair of the anterior cruciate ligament. Am J Sports Med. 1992;20(2):221–3.

Delvaux F, Rochcongar P, Bruyere O, et al. Return-to-play criteria after hamstring injury: actual medicine practice in professional soccer teams. J Sports Sci Med. 2014;13:721–3.

Ferretti A, Conteduca F, Morelli F, Masi V. Regeneration of the semitendinosus tendon after its use in anterior cruciate ligament reconstruction: a histologic study of three cases. Am J Sports Med. 2002;30(2):204–7.

Folsom GJ, Larson CM. Surgical treatment of acute versus chronic complete proximal hamstring ruptures: results of a new allograft technique for chronic reconstructions. Am J Sports Med. 2008 Jan;36(1):104–9.

Garrett WE Jr, Califf JC, Bassett FH III. Histochemical correlates of hamstring injuries. Am J Sports Med. 1984;12(2):98–103.

Gidwani S, Jagiello J, Bircher M. Avulsion fracture of the ischial tuberosity in adolescents--an easily missed diagnosis. BMJ. 2004 Jul 10;329(7457):99–100.

Glazer DD. Development and preliminary validation of the injury- psychological readiness to return to sport (I-PRRS) scale. J Athl Train. 2009;44:185–9.

Goldblatt JP, Fitzsimmons SE, Balk E, Richmond JC. Reconstruction of the anterior cruciate ligament: meta-analysis of patellar tendon versus hamstring tendon autograft. Arthroscopy. 2005 Jul;21(7):791–803.

Gregor RJ, Cavanagh PR, LaFortune M. Knee flexor moments during propulsion in cycling- a creative solution to Lombard's paradox. J Biomech. 1985;18(5):307–16.

Hachana Y, Chaabene H, Nabli MA, et al. Test-retest reliability, criterion-related validity, and minimal detectable change of the Illinois agility test in male team sport athletes. J Strength Cond Res. 2013;27:2752–9.

Hägglund M, Walden M, Ekstrand J. Risk factors for lower extremity muscle injury in professional soccer: the UEFA injury study. Am J Sports Med. 2013b;41:327–35.

Hägglund M, Walden M, Magnusson H, et al. Injuries affect team performance negatively in professional football: an 11-year followup of the UEFA champions league injury study. Br J Sports Med. 2013a;47:738–42.

Haines S, Baker T, Donaldson M. Development of a physical performance assessment checklist for athletes who sustained a lower extremity injury in preparation for return to sport: a Delphi study. Int J Sports Phys Ther. 2013;8:44–53.

Hawkins RD, Hulse MA, Wilkinson C, Hodson A, Gibson M. The association football medical research programme: an audit of injuries in professional football. Br J Sports Med. 2001;35(1):43–7.

Heck H, Mader A, Hess G, Mucke S, Muller R, Hollmann W. Justification of the 4-mmol/l lactate threshold. Int J Sports Med. 1985;6:117–30.

Heiderscheit BC, Sherry MA, Silder A, Chumanov ES, Thelen DG. Hamstring strain injuries: recommendations for diagnosis, rehabilitation, and injury prevention. J Orthop Sports Phys Ther. 2010 Feb;40(2):67–81.

Kellis E, Ellinoudis A, Kofotolis N. Hamstring elongation quantified using ultrasonography during the straight leg raise test in individuals with low back pain. PM R. 2015;7:576–83.

Koulouris G, Connell D. Hamstring muscle complex: an imaging review. Radiographics. 2005 May-Jun;25(3): 571–86.

Kvist J. Rehabilitation following anterior cruciate ligament injury: current recommendations for sport participation. Sports Med. 2004;34:296–80.

Langford JL, Webster KE, Feller JA. A prospective longitudinal study to assess psychological changes following anterior cruciate ligament reconstruction surgery. Br J Sports Med. 2009;43:377–8.

Linklater JM, Hamilton B, Carmichael J, Orchard J, Wood DG. Hamstring injuries: anatomy, imaging, and intervention. Semin Musculoskelet Radiol. 2010 Jun;14(2):131–61.

Liu H, Zhang Y, Rang M, Li Q, Jiang Z, Xia J, Zhang M, Gu X, Zhao C. Avulsion fractures of the ischial tuberosity: Progress of injury, mechanism, clinical manifestations, imaging examination, diagnosis and differential diagnosis and treatment. Med Sci Monit. 2018 Dec 27;24:9406–12.

Lombard WP, Abbott FM. The mechanical effects produced by the contraction of individual muscles of the thigh of the frog. Am J Physiol. 1907;20:1–60.

Malliaropoulos N, Isinkaye T, Tsitas K, et al. Reinjury after acute posterior thigh muscle injuries in elite track and field athletes. Am J Sports Med. 2011;39:304–10.

McCarty EC, Ritchie P, Gill HS, et al. Shoulder instability: return to play. Clin Sports Med. 2004;23:335–51.

Morgan BE, Oberlander MA. An examination of injuries in major league soccer. The inaugural season. Am J Sports Med. 2001;29(4):426–30.

Negra Y, Chaabene H, Hammami M, et al. Agility in young athletes: is it a different ability from speed and power? J Strength Cond Res. 2017;31:727–35.

Raya MA, Gailey RS, Gaunaurd IA, et al. Comparison of three agility tests with male servicemembers: Edgren side step test, t-test, and Illinois agility test. J Rehabil Res Dev. 2013;50:951–60.

Reurink G, Goudswaard GJ, Tol JL, et al. MRI observations at return to play of clinically recovered hamstring injuries. Br J Sports Med. 2014;48:1370–6.

Rouissi M, Chtara M, Berriri A, et al. Asymmetry of the modified Illinois change of direction test impacts young elite soccer players' performance. Asian J Sports Med. 2016;7:e33598.

Sanfilippo JL, Silder A, Sherry MA, et al. Hamstring strength and morphology progression after return to sport from injury. Med Sci Sports Exerc. 2013;45:448–54.

Simonian PT, Harrison SD, Cooley VJ, Escabedo EM, Deneka DA, Larson RV. Assessment of morbidity of semitendinosus and gracilis tendon harvest for ACL reconstruction. Am J Knee Surg. 1997;10(2):54–9.

Volpi P, Bisciotti GN. Muscle injury in athlete. Cham: The Italian Consensus Conference guidelines, Springer Nature; 2019.

Witvrouw E, Danneels L, Asselman P, et al. Muscle flexibility as a risk factor for developing muscle injuries in male professional soccer players. Am J Sports Med. 2003;31:41–6.

Zambaldi M, Beasley I, Rushton A. Return to play criteria after hamstring muscle injury in professional football: a Delphi consensus study. Br J Sports Med. 2017;51:1221–6.

10 Return to Training and Return to Play Following Adductor Injury

10.1 Anatomical Description

The adductor muscle complex is formed by:

1. Adductor longus
2. Adductor brevis
3. Adductor magnus
4. Pectineus
5. Gracilis
6. External obturator

10.1.1 Adductor Longus

The long adductor muscle (AL) is the most superficial among the adductor muscles. Its shape is flat and triangular. The AL arises as a flat and narrow tendon from the anterior surface of the pubic bone medial to the pubic tubercle and is distally inserted on the aponeurosis into the linea aspera of the femur (Tuite et al. 1998). Proximally, it presents a poorly vascularized fibrocartilaginous enthesis (Koulouris 2008; Davis et al. 2012). It is innervated by the obturator nerve. Several anatomical studies show that the AL anterior origin is tendinous while the posterior origin is muscular (Strauss et al. 2007; El Hage et al. 2010; Davis et al. 2012). Indeed, at its origin, the AL is composed of ~38% tendon and ~ 62% muscle tissue. About ~1.0 cm from the origin, the tendon percentage is ~34%, while at ~2.0 cm from the origin, the tendon represents ~27% of the total cross section (Strauss et al. 2007) (Fig. 10.1). With its action, it adducts and extra-rotates the thigh, also participating in the thigh flexion on the pelvis. It is innervated by the anterior branch of the obturator nerve of the lumbar plexus (L2–L4). The AL shows a central tendon that enters into the muscle belly about 5.5–8 cm from its origin, continuing as intramuscular tendon. The total length of the AL intramuscular tendon is between 7 and 17 cm (Davis et al. 2012). Inside the adductor muscle complex, the AL is most often implicated in the adductor related groin pain syndrome (Weir et al. 2015; Bisciotti et al. 2016a, b; Serner et al. 2018). It has been hypothesized that such strong involvement in the etiology of adductor related groin pain syndrome may be due to both its poor entheseal blood supply (Davis et al. 2012) and the fact that AL proximal insertion bears strong mechanical stress during the transmission of the force generated by muscle contraction (Renström and Peterson 1980; Tuite et al. 1998).

10.1.2 The External Obturator

The external obturator muscle is a flat, triangular muscle that covers the outer surface of the anterior wall of the pelvis. It arises from the lateral surface of the hip bone, immediately around the external surface of the obturator membrane and from the surrounding bone. The fibers are medially orientated, passing first obliquely by the

G. N. Bisciotti et al., *Return-to-Play after Lower Limb Muscle Injury in Football*,
https://doi.org/10.1007/978-3-030-84950-4_10

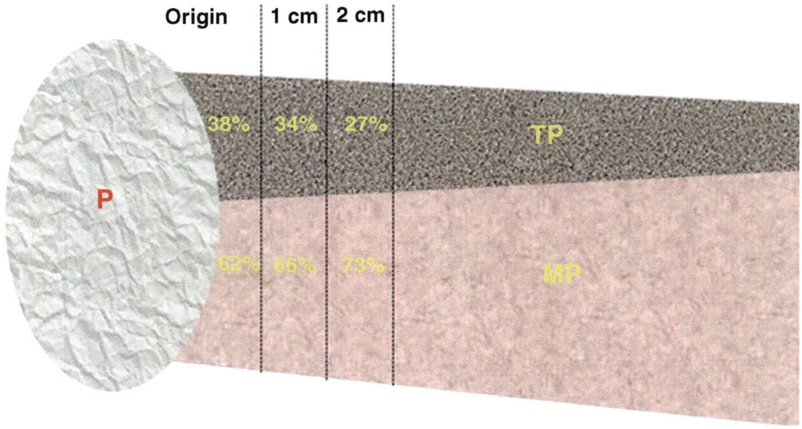

Fig. 10.1 AL sagittal schematic view. At its origin AL is composed of ~38% tendon and ~ 62% muscle tissue. At ~1.0 cm from the origin, the tendon percentage is ~34% and muscle tissue ~66%, while at ~2.0 cm from the origin, the tendon represents ~27% and the muscle tissue ~73%. Key: P = pubis bone; TP = tendinous part; M = muscle part

inferior and subsequently by the posterior surfaces of the femoral neck and then dorsally in the external obturator groove. Distally it inserts into the trochanteric fossa of the femur. Concerning the external obturator muscle, some anatomical variations may exist. A certain percentage of subjects (in average 33%) show a supernumerary muscle, located between the adductor brevis and the adductor minimus. When present, this muscle is similar to its neighboring adductors and is formed by separation from the superficial layer of the external obturator. For this reason, it is not ontogenetically related to the adductor hip muscle complex (Balius et al. 2018). The muscle is innervated by the obturator nerve. Its blood supply comes from the obturator artery. With its action it externally rotates the hip. Furthermore, it is an accessory adductor.

10.1.3 Adductor Brevis

The adductor brevis (AB) has a triangular shape and is placed in an inferior-posterior position with respect to the AL. It originates from the medial portion of the anterior aspect of the superior branch of the pubis and from the superior portion of the anterior aspect of the ischiopubic branch. Its distal insertion is at the level of the upper third of the medial lip of the linea aspera of the femur. Its action is to flex and extra-rotate the thigh. It is innervated by the anterior branch of the obturator nerve (L2–L4). Like the AL, also the AB shows an intramuscular tendon that covers about half of the entire muscular belly (Davis et al. 2012). Some authors have described how some fibers of the AB (muscle in man and tendon in women) merge with the gracilis muscle (Davis et al. 2012).

10.1.4 Adductor Magnus

The adductor magnus (AM) is a large triangle-shaped muscle situated on the medial side of the thigh. The muscle consists of two parts: one arises from the ischiopubic ramus (i.e., a small part of the inferior ramus of the pubis and the inferior ramus of the ischium) and is called the "pubofemoral portion" (also called "adductor portion" or "adductor minimus"), the other AM portion arises from the ischium tuberosity and is called the "ischiocondylar portion" (or also the "extensor portion" or "hamstring portion"). Due to their common embryonic origin, innervation, and action, the ischiocondylar portion (or hamstring portion) is often considered part of the hamstring muscle complex. The AM ischiocondylar portion is considered a muscle of the posterior compartment of the thigh, while the AM pubofemoral portion is considered a muscle of the medial compartment (Obey et al. 2016).

10.1.4.1 Pubofemoral (Adductor) Portion

The pubofemoral fibers arising from the pubis ramus are short and horizontal in direction. They insert into the rough line of the femur leading from the great trochanter to the linea aspera medial to the gluteus maximus. The pubofemoral fibers arising from the ramus of the ischium are directed downward and laterally with different degrees of obliquity. They insert by means of a broad aponeurosis into the linea aspera and the upper part of its medial prolongation below (Obey et al. 2016).

10.1.4.2 Ischiocondylar (Hamstring) Portion

The medial portion of the muscle is composed principally of the fibers arising from the tuberosity of the ischium. It forms a thick fleshy mass consisting of coarse bundles descending almost vertically and ending about the lower third of the thigh in a rounded tendon which is inserted into the adductor tubercle. This distal insertion is connected by a fibrous expansion to the line leading upward from the tubercle to the linea aspera.

The AM shows an anatomical particularity. Indeed, the upper, lateral part of the AM is an incompletely separated division often considered a separate muscle (i.e., the adductor minimus). The two muscles are frequently separated by a branch of the superior perforating branch of the profunda femoris artery.

Being a "composite muscle," the AM is innervated by two different nerves. The adductor portion is innervated by the posterior division of the obturator nerve, while the hamstring portion is innervated by the sciatic nerve (Takizawa et al. 2014). The perforating branches of the profunda femoris artery, passing through the AM, are the primary source of the blood supply to the muscle. As with the other adductor muscles, the medial femoral circumflex artery supplies the superior portion of the muscle.

The AM muscle is a powerful thigh adductor and is active when the legs are moved from a wide spread position to one in which the legs are parallel to each other. The part attached to the linea aspera acts as a lateral rotator. The part reaching the medial epicondyle acts as a medial rotator when the leg is rotated outward and flexed, acting also to extend the hip joint.

10.1.5 Pectineus

The pectineus muscle (from the Latin word *pecten*, meaning comb) is a flat, quadrangular muscle that is located in the anterior part of the thigh upper and medial aspect. The pectineus is the most anterior adductor muscle of the hip. The muscle arises from the pectineal line of the pubis, from a little portion of bone between the iliopectineal eminence and pubic tubercle and from the fascia covering the anterior surface of the muscle. Its fibers insert into the pectineal line of the femur (Attum and Varacallo 2020).

Also the pectineus muscle, like the adductor magnus, is considered a "composite muscle," and its innervation is by the femoral nerve (L2 and L3) and occasionally (20% of the subjects) from a branch of the obturator nerve (the so called "accessory obturator nerve"). If the accessory obturator nerve is present, it innervates a portion of the pectineus muscle, entering the muscle on its dorsomedial aspect. In any case, the main nerve of the muscle is the femoral nerve, which is always present and provides the sole innervation for the pectineus muscle in over 90% of cases. The superficial part of the muscle is supplied by the medial circumflex femoral artery, a branch of the femoral artery. The deep portion of the muscle is vascularized by the anterior branch of obturator artery that is a branch of the internal iliac artery (Lytle 1979).

The muscle primary function is hip flexion, but it also adducts and internally rotates the thigh.

10.1.6 Gracilis Muscle

The gracilis muscle (from the Latin word "*gracilis*" meaning "slender") is the most superficial muscle on the medial aspect of the thigh. Its shape is thin and flattened, broad above, narrow, and tapering below. The muscle arises by a thin aponeurosis from the anterior margins of the lower half of the symphysis pubis and the upper half of the pubic arch. Then the muscle's fibers run vertically downward, ending in a rounded shaped tendon. The tendon passes behind the femur medial condyle after curves around the tibial medial condyle where it becomes flattened. Finally, it inserts into the upper part of the medial surface of the body of the tibia, below the condyle. At its distal insertion the tendon is situated immediately above that of the semitendinosus muscle, and its upper edge is overlapped by the tendon of the sartorius muscle which it joins to form the pes anserinus. A few of the fibers of the lower part of the tendon are prolonged into the deep fascia of the leg (Dziedzic et al. 2018). The gracilis muscle is used in ACL reconstruction (Pasquini et al. 2017). Furthermore, gracilis muscle is widely used in reconstructive surgery (graciloplasty), either as a pedicled flap or as a free microsurgical flap. As a functioning pedicled flap, the gracilis muscle can be transferred for the treatment of anal incontinence. This technique called "graciloplasty" was described in the 1950s by Pickrell (Seccia et al. 2001).

The gracilis muscle is innervated by the obturator nerve. The blood supply derives from the medial circumflex femoral artery.

The gracilis adducts, medially rotates (with hip flexion), laterally rotates, and flexes the hip. The muscle also aids in flexion of the knee.

10.2 Epidemiological Notes

Adductors' injuries are a common problem in professional soccer, and they accounted for 23% of the muscle injuries, resulting in 14 days of absence from sports (Ekstrand et al. 2011). Groin strains have the highest incidence in the age group between 22 and 30 years. The reinjury rate after adductor muscle injuries (either acute or chronic) has been reported as high as 18% (Ekstrand et al. 2011). Previous injury and a history of reduced adductor muscle strength have been identified as risk factors for adductor muscle injuries (Engebretsen et al. 2008; Engebretsen et al. 2010). Soccer players with a previous groin injury are twice as likely to develop a new groin strain (Engebretsen et al. 2010). Interestingly, the elite ice hockey players whose preseason adductor strength is 80% or less of the abductor strength have been identified as having 17 times a higher risk of suffering an adductor injury during the season (Tyler et al. 2002). Amateur soccer players with weak adductors had four times as much risk for a new groin injury (Engebretsen et al. 2010). Preseason strengthening has been effective in reducing the amount of adductor muscle injuries in hockey players during the season (Tyler et al. 2002).

Adductor tendinopathy is a very common clinical situation in etiopathogenesis of groin pain syndrome (GPS) in several sports, often reducing the players' activity when chronic (Karlsson et al. 2014; Weir et al. 2015). Indeed, its yearly incidence in football (soccer) of 10–18% continues to increase due to many risk factors such as high loads and short recoveries (Mosler et al. 2018). It is important to note that the majority of injury surveillance studies in football are based on the so-called "time loss concept" (Werner et al. 2009). Indeed, injuries are recorded only if a player is unable to participate in football training and/or competition (Junge and Dvorak 2004; Waldén et al. 2007; Dvorak et al. 2011; Bjørneboe et al. 2014; Noya Salces et al. 2014). Recent studies revealed that the "time loss definition" captured only one-third of the GPS injuries in male football players (Harøy et al. 2017). Thus, the traditional "time loss injury" approach may be inappropriate, and the recorded data may represent only the "tip of the iceberg" of a deeper and most painful issue (Harøy et al. 2017). Indeed, it is common for players to continue training despite the pain, in order not to register any time loss injury, yet by doing so, the affected structures are subjected to overuse, which is consolidated as an important element in most cases of GPS (Bahr 2009¸Werner et al. 2019). For adductor-related GPS, overuse etiology is confirmed by a recent study in which football adductor injuries are 27% traumatic and 73% due to overuse (Werner et al. 2019).

10.3 Clinical and Imaging Assessments for RTT

The tests suggested by the CC following SGMC indirect injury are the following:

10.3.1 General Assessment

Below are the Consensus Conference recommendations for clinical and imaging assessments for RTT following adductor injury:

1. Absence of clinical symptoms (Kvist 2004; Malliaropoulos et al. 2011; Delvaux et al. 2014).
2. Absence of pain or tenderness during muscle palpation (Kvist 2004; Zambaldi et al. 2017; Bisciotti and Volpi 2018; Reurink et al. 2014).
3. Absence of pain on passive and active stretching. (Witvrouw et al. 2003; Bisciotti 2015; Bisciotti and Volpi 2018).
4. Absence of pain on isometric, concentric, and eccentric contraction (Bisciotti and Volpi 2018).
5. Completion of the prescribed rehabilitation program (Reurink et al. 2014).
6. MRI and US imaging assessment points will be specified in Chap. 5: "The Role of Imaging in the Return to Training and Return to Play Decision-Making Process."
7. Subjective feelings of the player taken into account (i.e., assess levels of anxiety, apprehension, fear of failure, and/or fear of reinjury) (McCarty et al. 2004; Bauman 2005; Glazer 2009; Langford et al. 2009; Clover and Wall 2010; Delvaux et al. 2014).

10.3.2 Specific Assessment

The specific assessment for adductors muscles indirect injury is based on the following tests:

1. Pubic stress test (Hogan and Lovell 1998; Bisciotti 2013; Bisciotti et al. 2016a, b).
2. Resisted hip adduction test (Hogan et al., Lovell, 1998; Croisier et al. 2002; Engebretsen et al. 2010).
3. Squeeze test (Delahunt et al. 2011; Nevin and Delahunt 2014; Bisciotti et al. 2016a, b).
4. Adductor passive stretching test (Witvrouw et al. 2003; Atkinson et al. 2010).

10.4 Laboratory Tests for RTT

Below are the recommended laboratory tests for RTT following adductor injury:

1. Adductor muscle strength assessed by dynamometric tests (Croisier et al. 2002; Sanfilippo et al. 2013; Delvaux et al. 2014; Bisciotti et al. 2016a, b).

10.5 Field Tests for RTT

The following are the field tests recommended by the CC prior to RTT after adductor muscle injuries:

1. Kicking test (Bisciotti et al. 2016a, b).
2. Carioca test (Kong et al. 2012; Jang et al. 2014).
3. Agility T-test (Sassi et al. 2009.

10.6 RTP Tests after Adductor Injury

As in the case of quadriceps and hamstring injuries, the criteria for the RTP-DMP are related to performance evaluation. For this reason, the RTP-DMP must chronologically follow the RTT-DMP. Therefore, only if the athlete has obtained a positive judgment for the RTT, the period of registration of the tests for RTP begins.

In this regard, the CC recommend the following points:

1. The data acquisition period must start from the first day of RTP and last 7–10 days.
2. During this period, the performance data must be systematically recorded via the GPS system.
3. It is necessary to identify some "typical" sessions, substantially overlapping each other, which can be deduced from the last pre-injury period.

As already mentioned in Chap. 4, the three evaluation categories are:

1. Quantitative evaluation.
2. Qualitative evaluation.
3. Parameter analysis.

All the indices described in Chap. 4 as related to the abovementioned categories must be analyzed. The reference value below which the positive judgment for RTP is postponed is set at a maximum difference of 10% between the pre-injury data and the data recorded during the acquisition period following the RTT. Concerning the control of aerobic fitness, the CC experts suggest that the RTP-DMP be implemented by means of a test for VO_2max determination (Heck et al. 1985; Chamari et al. 2004; Chamari et al. 2005; Badawy and Muaidi 2018), in which the player must record a value of VO_2max, and consequently of VAM, equal to at least 90% of the pre-injury values.

References

Atkinson HDE, Johal P, Falworth MS, et al. Adductor tenotomy: its role in the management of sports-related chronic groin pain. Arch Orthop Trauma Surg. 2010;130:965–70.

Attum B, Varacallo M. Anatomy, bony pelvis and lower limb, thigh muscles. In: StatPearls [Internet]. Treasure Island: StatPearls Publishing; 2020 Jan.

Badawy MM, Muaidi QI. Aerobic profile during high-intensity performance in professional Saudi Athletes. Pak J Biol Sci. 2018;21(1):24–8.

Bahr R. No injuries, but plenty of pain? On the methodology for recording overuse symptoms in sports. Br J Sports Med. 2009 Dec;43(13):966–72.

Balius R, Susín A, Morros C, Pujol M, Pérez-Cuenca D, Sala-Blanch X. Gemelli-obturator complex in the deep gluteal space: an anatomic and dynamic study. Skelet Radiol. 2018 Jun;47(6):763–70.

Bauman J. Returning to play: the mind does matter. Clin J Sport Med. 2005;15:432–5.

Bisciotti GN. La tendinopatia degli adduttoria nel calciatore quando Il ritorno alla corsa? Strength Condition. 2013;5:11–6.

Bisciotti GN. Return to play after a muscle lesion. In: Volpi P, editor. Arthroscopy in sport. Cham: Springer Edition; 2015.

Bisciotti GN, Quaglia A, Belli A, et al. Return to sports after ACL reconstruction: a new functional test protocol. Muscl Ligam Tendons J. 2016a;06:499–509.

Bisciotti GN, Volpi P, Zini R, et al. Groin Pain Syndrome Italian Consensus Conference on terminology, clinical evaluation and imaging assessment in groin pain in athlete. BMJ Open Sport Exerc Med. 2016b;2(1):e000142.

Bisciotti GN, Volpi P. Return to play. In: Volpi P, editor. Football doctor manual. Trento: Edra Edition; 2018. p. 247–59.

Bjørneboe J, Bahr R, Andersen TE. Gradual increase in the risk of match injury in Norwegian male professional football: a 6-year prospective study. Scand J Med Sci Sports. 2014;24(1):189–96.

Chamari K, Hachana Y, Ahmed YB, Galy O, Sghaïer F, Chatard JC, Hue O, Wisløff U. Field and laboratory testing in young elite soccer players. Br J Sports Med. 2004 Apr;38(2):191–6.

Chamari K, Moussa-Charai I, Boussaïdi L, Hachana Y, Kauech F, Wisløff U. Appropriate interpretation of aerobic capacity: allometric scaling in adult and young soccer players. Br J Sports Med. 2005 Feb;39(2):97–101.

Clover J, Wall J. Return-to-play criteria following sports injury. Clin Sports Med. 2010;29:169–75.

Croisier J-L, Forthomme B, Namurois M-H, et al. Hamstring muscle strain recurrence and strength performance disorders. Am J Sports Med. 2002;30:199–203.

Davis JA, Stringer MD, Woodley SJ. New insights into the proximal tendons of adductor longus, adductor brevis and gracilis. Br J Sports Med. 2012;46(12):871–6.

Delahunt E, McEntee BL, Kennelly C, et al. Intrarater reliability of the adductor squeeze test in gaelic games athletes. J Athl Train. 2011;46:241–5.

Delvaux F, Rochcongar P, Bruyere O, et al. Return-to-play criteria after hamstring injury: actual medicine practice in professional soccer teams. J Sports Sci Med. 2014;13:721–3.

Dvorak J, Junge A, Derman W, Schwellnus M. Injuries and illnesses of football players during the 2010 FIFA World Cup. Br J Sports Med. 2011 Jun;45(8):626–30.

Dziedzic DW, Bogacka U, Komarnițki I, Ciszek B. Anatomy and morphometry of the distal gracilis muscle tendon in adults and foetuses. Folia Morphol (Warsz). 2018;77(1):138–43.

Ekstrand J, Hagglund M, Walden M. Epidemiology of muscle injuries in professional football (soccer). Am J Sports Med. 2011;39:12261232.

El Hage S, Rachkidi R, Noun Z, Haidar R, Dagher F, Kharrat K, Ghanem I. Is percutaneous adductor tenotomy as effective and safe as the open procedure? J Pediatr Orthop. 2010 Jul-Aug;30(5):485–8.

Engebretsen A, Myklebust G, Holme I, Engebretsen L, Bahr R. Intrinsic risk factors for groin injuries among male soccer players: a prospective cohort study. Am J Sports Med. 2010;38:2051–7.

Engebretsen AH, Myklebust G, Holme I, Engebretsen L, Bahr R. Prevention of injuries among male soccer players: a prospective, randomized intervention study targeting players with previous injuries or reduced function. Am J Sports Med. 2008;36:1052–60.

Glazer DD. Development and preliminary validation of the injury- psychological readiness to return to sport (I-PRRS) scale. J Athl Train. 2009;44:185–9.

Harøy J, Andersen TE, Bahr R. Groin problems in male soccer players are more common than previously reported: response. Am J Sports Med. 2017 Nov;45(13):NP32–3.

Heck H, Mader A, Hess G, Mucke S, Muller R, Hollmann W. Justification of the 4-mmol/l lactate threshold. Int J Sports Med. 1985;6:117–30.

Hogan A, Lovell G. The groin pain provocation test. In: Brown A, editor. 4thWorld Football Symposium Conference Proceedings. London: Routledge; 1998.

Jang SH, Kim JG, Ha JK, et al. Functional performance tests as indicators of returning to sports after anterior cruciate ligament reconstruction. Knee. 2014;21: 95–101.

Junge A, Dvorak J. Soccer injuries: a review on incidence and prevention. Sports Med (Auckland, NZ). 2004;34(13):929–38.

Karlsson MK, Dahan R, Magnusson H, Nyquist F, Rosengren BE. Groin pain and soccer players: male versus female occurrence. J Sports Med Phys Fitness. 2014 Aug;54(4):487–93.

Kong DH, Yang SJ, Ha JK, et al. Validation of functional performance tests after anterior cruciate ligament reconstruction. Knee Surg Relat Res. 2012;24: 40–5.

Koulouris G. Imaging review of groin pain in elite athletes: an anatomic approach to imaging findings. AJR Am J Roentgenol. 2008;191:962–72.

Kvist J. Rehabilitation following anterior cruciate ligament injury: current recommendations for sport participation. Sports Med. 2004;34:296–80.

Langford JL, Webster KE, Feller JA. A prospective longitudinal study to assess psychological changes following anterior cruciate ligament reconstruction surgery. Br J Sports Med. 2009;43:377–8.

Lytle WJJ. Inguinal anatomy. Anatolia. 1979 May;128(Pt 3):581–94.

Malliaropoulos N, Isinkaye T, Tsitas K, et al. Reinjury after acute posterior thigh muscle injuries in elite track and field athletes. Am J Sports Med. 2011;39(2): 304–10.

McCarty EC, Ritchie P, Gill HS, et al. Shoulder instability: return to play. Clin Sports Med. 2004;23:335–51.

Mosler AB, Weir A, Eirale C, Farooq A, Thorborg K, Whiteley RJ, Hölmich P, Crossley KM. Epidemiology of time loss groin injuries in a men's professional football league: a 2-year prospective study of 17 clubs and 606 players. Br J Sports Med. 2018 Mar;52(5):292–7.

Nevin F, Delahunt E. Adductor squeeze test values and hip joint range of motion in Gaelic football athletes with longstanding groin pain. J Sci Med Sport. 2014;17:155–9.

Noya Salces J, Gomez-Carmona PM, Gracia-Marco L, Moliner-Urdiales D, Sillero-Quintana M. Epidemiology of injuries in first division Spanish football. J Sports Sci. 2014;32(13):1263–70.

Obey MR, Broski SM, Spinner RJ, Collins MS, Krych AJ. Anatomy of the Adductor Magnus Origin: Implications for Proximal Hamstring Injuries. Orthop J Sports Med. 2016 Jan 11;4(1):2325967115625055.

Pasquini A, Jacopetti M, Pogliacomi F, Ramazzina I, Costantino C. Neuromuscular recovery in ACL reconstruction with Bone-Tendon-Patellar-Bone and Semitendinosus-Gracilis autograft. Acta Biomed. 2017 Oct 18;88(4S):62–8.

Renström P, Peterson L. Groin injuries in athletes. Br J Sports Med. 1980;14:30–6.

Reurink G, Goudswaard GJ, Tol JL, et al. MRI observations at return to play of clinically recovered hamstring injuries. Br J Sports Med. 2014;48(18):1370–6.

Sanfilippo JL, Silder A, Sherry MA, et al. Hamstring strength and morphology progression after return to sport from injury. Med Sci Sports Exerc. 2013;45: 448–54.

Sassi RH, Dardouri W, Yahmed MH, Gmada N, Mahfoudhi ME, Gharbi Z. Relative and absolute reliability of a modified agility T-test and its relationship with vertical jump and straight sprint. J Strength Cond Res. 2009 Sep;23(6):1644–51.

Seccia M, Banti P, Zocco G, Viacava P. Restoration of fecal continence with chronic electrostimulation of gracilis muscle 17 years after a Pickrell's operation. Int J Color Dis. 2001 Nov;16(6):391–4.

Serner A, Weir A, Tol JL, Thorborg K, Roemer F, Guermazi A, Yamashiro E, Hölmich P. Characteristics of acute groin injuries in the adductor muscles: A detailed MRI study in athletes. Scand J Med Sci Sports. 2018 Feb;28(2):667–76.

Strauss EJ, Campbell K, Bosco JA. Analysis of the cross-sectional area of the adductor longus tendon: a descriptive anatomic study. Am J Sports Med. 2007;35:996–9.

Takizawa M, Suzuki D, Ito H, Fujimiya M, Uchiyama E. The adductor part of the adductor magnus is innervated by both obturator and sciatic nerves. Clin Anat. 2014 Jul;27(5):778–82.

Tuite DJ, Finegan PJ, Saliaris AP, Renström PA, Donne B, O'Brien M. Anatomy of the proximal musculotendinous junction of the adductor longus muscle. Knee Surg Sports Traumatol Arthrosc. 1998;6(2):134–7.

Tyler TF, Nicholas SJ, Campbell RJ, Donellan S, McHugh MP. The effectiveness of a preseason exercise program to prevent adductor muscle strains in professional ice hockey players. Am J Sports Med. 2002;30:680–3.

Waldén M, Hägglund M, Ekstrand J. Football injuries during European championships 2004-2005. Knee Surg Sports Traumatol Arthrosc. 2007 Sep;15(9): 1155–62.

Weir A, Brukner P, Delahunt E, Ekstrand J, Griffin D, et al. Doha agreement meeting on terminology and definitions in groin pain in athletes. Br J Sports Med. 2015 Jun;49(12):768–74.

Werner J, Hägglund M, Ekstrand J, Waldén M. Br J Sports Med. 2019 May;53(9):539–46.
Werner J, Hagglund M, Walden M, Ekstrand J. UEFA injury study: a prospective study of hip and groin injuries in professional football over seven consecutive seasons. Br J Sports Med. 2009;43(13):1036–40.
Witvrouw E, Danneels L, Asselman P, et al. Muscle flexibility as a risk factor for developing muscle injuries in male professional soccer players. Am J Sports Med. 2003;31:41–6.
Zambaldi M, Beasley I, Rushton A. Return to play criteria after hamstring muscle injury in professional football: a Delphi consensus study. Br J Sports Med. 2017;51:1221–6.

11 Return to Training and Return to Play Following Soleus-Gastrocnemius Injury

11.1 Anatomical Description

The soleus muscle and the medial and lateral gastrocnemius muscles compose, with the posterior tibial muscle, the so-called posterior compartment of the leg.

11.1.1 Soleus Muscle

The soleus muscle is located behind the gastrocnemius muscle, in the superficial posterior compartment of the leg. It is a large and flat muscle. Unlike gastrocnemius, the soleus consists of a single muscle head, which is very large at the level of its origin. It originates from the back of the fibula head and from an oblique bone crest located on the medial margin of the tibia, which is known by the name of "soleal line." Distally, it inserts, via the Achilles tendon, on the heel. The soleus exhibits significant morphological differences across species, and it is unipennate in many of them. In the human species, soleus is a complex multi-pennate muscle, usually showing a separate posterior aponeurosis from the gastrocnemius muscle. When precisely inspecting its conformation, the following can be observed:

1. The majority of its fibers originate from each side of the anterior aponeurosis (Agur et al. 2003; Hodgson et al. 2006).
2. Other fibers originate from the posterior surfaces of the head of the fibula, while its upper quarter and middle third originate from the medial border of the tibia.
3. The fibers originating from the anterior surface of the anterior aponeurosis insert onto the median septum, while the fibers originating from the posterior surface of the anterior aponeurosis insert onto the posterior aponeurosis.
4. The posterior aponeurosis and median septum join in the lower quarter of the muscle and then join with the anterior aponeurosis of the gastrocnemius muscles to form the Achilles tendon.

In some species of animals, such as the rabbit, the soleus is fused for much of its length with the gastrocnemius muscle. On the contrary, in humans, soleus and gastrocnemius muscles are relatively separate, so much that shear can be detected between the soleus and gastrocnemius aponeurosis (Bojsen-Møller et al. 2004). Remaining in the field of comparative anatomy, it is interesting to note that the soleus is only vestigial in the horse (Meyers and Hermanson 2006). Back to humans, the plantaris muscle, together with a portion of its tendon, run between the soleus and gastrocnemius muscles. The transverse intermuscular septum is located more in deep, separating the superficial posterior compartment of the leg from the deep posterior

G. N. Bisciotti et al., *Return-to-Play after Lower Limb Muscle Injury in Football*,
https://doi.org/10.1007/978-3-030-84950-4_11

compartment. The soleus is innervated by the tibial nerve, and its blood supply comes from the popliteal artery (which is a derivation of the femoral artery), the peroneal artery, and the posterior tibial artery. The soleus has a high percentage of slow muscle fibers, higher than many other muscles. In some animals, such as the guinea pig and the cat, the soleus has 100% of slow muscle fibers (Burke et al. 1974; Ariano et al. 1972). The human soleus fiber composition is quite variable, containing between 60 and 100% of slow fibers (Gollnick et al. 1974). During walking, the soleus is the primary muscle utilized for plantarflexion due to its high percentage of slow-twitch fibers resisting fatigue (Saladin 2007).

The soleus action is the foot plantarflexion. It is the most effective muscle for plantarflexion in a bent knee position, which is why it is also called the "first gear muscle." The soleus plays an important role in maintaining the standing posture: with its constant pull it prevents the body from falling forward. Furthermore, in the upright posture, the soleus is responsible for pumping venous blood back into the heart from the periphery. For this reason, it is often called the "skeletal muscle pump," the "peripheral heart," or the "tricipital pump" (Botta et al. 2001).

11.1.2 Gastrocnemius Muscle

The gastrocnemius muscle is a superficial two-headed muscle running from just above the knee to the heel. It is a three-joint muscle, intersecting the knee, the ankle, and the subtalar joint. The name of this muscle is derived from Greek words *γαστήρ* (belly or stomach) and *κνήμη* (leg), thus meaning "stomach of the leg," referring to the bulging shape of the calf. The lateral and medial heads originate from the lateral and medial condyle of the femur, respectively. Distally it forms a common tendon with the soleus muscle, known as the calcaneal tendon or Achilles tendon, and inserts onto the posterior surface of the calcaneus. Since they share a common insertion via the Achilles tendon, some anatomists consider the soleus and gastrocnemius as a single muscle (i.e., the triceps surae). In average, 10–30% of subjects show a sesamoid bone, called "fabella" in the lateral (outer) head of the gastrocnemius muscle. It is innervated by the tibial nerve (L4-S1). The gastrocnemius medial and lateral heads are supplied by the medial and lateral sural artery, respectively, and by branches of the popliteal artery. Additionally, anastomotic vessels cross between the two bellies. Its function is the plantar flexion of the foot at the ankle joint and flexing the leg at the knee joint. The gastrocnemius is primarily involved in running, jumping, and other "fast" movements of the leg and to a lesser degree in walking and standing (Hamilton and Luttgens 2001). Unlike the soleus, the gastrocnemius has a high percentage of fast-twitch fibers (Hamilton and Luttgens 2001). This characteristic justifies its role during movements performed at high contraction speeds (Saladin 2007).

11.2 Epidemiological Notes

The soleus-gastrocnemius muscle complex injuries (SGMC) are a relatively common traumatic event in many sports activities in which the training and competition workload is mainly based on speed run, acceleration, and deceleration phases and especially when the play conditions are strenuous (Orchard 2001; Orchard and Seward 2014; Green and Pizzari 2017). In soccer the SGMC injuries are frequent, reaching being on average 0.84 cases every 1000 exposure hours (Carling et al. 2011); also in rugby the SGMC injuries are very frequent with a risk ratio that is between 0.98 and 5.85 (Orchard 2001). In Australian football SGMC injuries are one of the most frequent traumatic events (three injuries per team per season), with a high degree of reinjury risk which is on average equal to 16% (Orchard and Seward 2014). SGMC injuries are also frequent in pole vault (Rebella 2015), professional dance (Luke et al. 2002), triathlon (Korkia et al. 1994), and tennis (Colberg et al. 2015). Some authors showed that in the case of recurrence the SGMC injuries require a rehabilitation period relatively long (Carling et al. 2011). It is important to note that the SGMC injuries mainly occur during some critical periods of the season, like at the end of competition period (Mallo et al. 2011). Older players (above 25.8 ± 4.5 years) have a risk at least two times greater of incurring in SGMC injuries (Hagglund et al. 2013a, b). In literature, no similar association was found for other muscle groups (Hagglund et al. 2013a, b). The risk factors showing greater evidence for SGMC injuries are a previous history of injuries in the same anatomical zone and age (Hagglund et al. 2013a, b). On the contrary, previous injuries in other muscle groups (hamstring, quadriceps and adductors) show only limited evidence for the association with the SGMC injuries (Hagglund et al. 2013a, b). Furthermore, there is no evidence for other factors, such as height, weight, gender, and limb dominance, to be risk factors for SGMC injuries (Hagglund et al. 2013a, b).

It is important to note that unlike the gastrocnemius, the soleus is considered to be at a low risk for injury. The reason is that it crosses only one joint (i.e., the ankle) and it is composed primarily of slow-twitch fibers (type I). Furthermore the soleus is predominantly involved in postural control and low speed activity, and it is damaged while the knee joint is flexed (Koulouris et al. 2007; Balius et al. 2014). Unfortunately, for those reasons, the soleus injuries are frequently underestimated, resulting in a delayed diagnosis (Lundgren and Davis 2004). Just as unfortunately the use of US examination may underdiagnose soleus injuries because the muscle is more deeply located than the gastrocnemius, is a multipennate muscle, and is importantly vascularized (Balius et al. 2014). For these reasons MRI has become the gold standard imaging in the assessment of soleus injuries (Pezzotta et al. 2017).

The objective difficulty of the US soleus examination and the justification for the choice of the MRI investigation can be summarized in the following points:

1. The soleus is a muscle presenting an anatomical complexity. In particular, it shows the presence of three tendon aponeuroses with a well-represented venous vascularization.
2. The muscle is deep located and often hypertrophic in professional athletes. These characteristics determine a higher US attenuation.
3. The soleus connective tissue structure is characterized by variable pennation angles that are responsible for anisotropy artifacts.
4. The soleus injuries often show a subacute presentation, due to the higher concentration of slow fibers.

We would like to underline that a soleus injury early diagnosis is extremely important especially in professional players, because a delay in diagnosis could lead the player to continue sport activity, increasing the risk for injury aggravation and/or reinjury.

Finally, it is important to note that the soleus injuries located in the central tendon or muscular fascia have a worse prognosis (Balius et al. 2013).

11.3 Clinical and Imaging Assessments for RTT

11.3.1 General Assessment

Below are the Consensus Conference recommendations for clinical and imaging assessments for RTT following soleus-gastrocnemius muscle complex injury:

1. Absence of clinical symptoms (Kvist 2004; Malliaropoulos et al. 2011; Delvaux et al. 2014).
2. Absence of pain or tenderness during muscle palpation (Kvist 2004; Zambaldi et al. 2017; Bisciotti and Volpi 2018; Reurink et al. 2014).
3. Absence of pain on passive and active stretching (Witvrouw et al. 2003; Bisciotti 2015; Bisciotti and Volpi 2018).
4. Absence of pain on isometric, concentric, and eccentric contraction (Bisciotti and Volpi 2018).
5. Completion of the prescribed rehabilitation program (Reurink et al. 2014).
6. MRI and US imaging assessment points will be specified in Chap. 5: "The Role of Imaging in the Return to Training and Return to Play Decision-Making Process."

7. Subjective feelings of the player taken into account (i.e., assess levels of anxiety, apprehension, fear of failure, and/or fear of reinjury) (McCarty et al. 2004; Bauman 2005; Glazer 2009; Langford et al. 2009; Clover and Wall 2010; Delvaux et al. 2014).

11.3.2 Specific Assessment

The specific assessment for SGMC injuries is based on the following tests:

1. Heel-raise test (Moller et al. 2005; Harris-Love et al. 2014).
2. Weight bearing lunge test (Hoch and McKeon 2011; Powden et al. 2015; Baumach et al. 2016).

11.4 Laboratory Tests for RTT

The following are the recommended laboratory tests prior to RTT recommended by the CC after SGMC injuries:

1. Soleus-gastrocnemius muscle strength assessed by dynamometric tests (Croisier et al. 2002; Malliaropoulos et al. 2011; Askling et al. 2010; Sanfilippo et al. 2013).
2. Synchro plates test (Bisciotti et al. 2016).
3. Drop jump test (Silbernagel et al. 2006; Hewett et al. 2007; Powell et al. 2018).

11.5 Field Tests for RTT

The following are the recommended field tests prior to RTT recommended by the CC after SGMC injuries:

1. Illinois Agility Test (Hachana et al. 2013; Raya et al. 2013; Bisciotti 2015; Negra et al. 2017; Bisciotti and Volpi 2018).
2. Agility T-test (Sassi et al. 2009).

11.6 RTP Tests After SGMC Injury

As in the case of quadriceps, hamstring, and adductor injuries, the criteria for the RTP-DMP are related to the evaluation of the football performance. Therefore the RTP-DMP will chronologically follow the RTT-DMP only if the athlete has obtained a positive judgment for the RTT. In such case, the period of registration of the tests for RTP begins.

In this regard, the CC recommend the following points:

1. The data acquisition period must start from the first day of RTP and last 7–10 days.
2. During this period the performance data must be systematically recorded via the GPS system.
3. It is necessary to identify some "typical" sessions, substantially overlapping each other, which can be deduced both from the last pre-lesional week and from the post-period.

As already mentioned in Chap. 4, the three evaluation categories are:

1. Quantitative evaluation.
2. Qualitative evaluation.
3. Parameter analysis.

All the indices described in Chap. 4 as related to the abovementioned categories must be analyzed. The reference value below which the positive judgment for RTP is postponed is set at a maximum difference of 10% between the pre-lesional data and the data recorded during the acquisition period following the RTT. For SGMC injuries too, concerning the control of aerobic fitness, the CC experts suggest that the RTP-DMP be implemented by means of a test for VO_2max determination (Heck et al. 1985; Chamari et al. 2004; Chamari et al. 2005; Badawy and Muaidi 2018), in which the player must record a value of VO_2max, and consequently of VAM, equal to at least 90% of the pre-injury values.

References

Agur AM, Ng-Thow-Hing V, Ball KA, Fiume E, McKee NH. Documentation and three-dimensional modelling of human soleus muscle architecture. Clin Anat. 2003;16(4):285–93.

Ariano MA, Armstrong RB, Edgerton VR. Hindlimb muscle fiber populations of five mammals. J Histochem Cytochem. 1972;21(1):51–5.

Askling CM, Nilsson J, Thorstensson A. A new hamstring test to complement the common clinical examination before return to sport after injury. Knee Surg Sports Traumatol Arthrosc. 2010;18:1798–803.

Badawy MM, Muaidi QI. aerobic profile during high-intensity performance in professional Saudi Athletes. Pak J Biol Sci. 2018;21(1):24–8.

Balius R, Alomar X, Rodas G, et al. The soleus muscle: MRI, anatomic and histologic findings in cadavers with clinical correlation of strain injury distribution. Skelet Radiol. 2013;42(4):521–30.

Balius R, Rodas G, Pedret C, et al. Soleus muscle injury: sensitivity of ultrasound patterns. Skelet Radiol. 2014;43(6):805–12.

Baumach SF, Braunstein M, Regauer M, et al. Diagnosis of musculus gastrocnemius tightness - key factors for the clinical examination. J Vis Exp. 2016;113

Bauman J. Returning to play: the mind does matter. Clin J Sport Med. 2005;15:432–5.

Bisciotti GN. Return to play after a muscle lesion. In: Volpi P, editor. Arthroscopy in sport. Cham: Springer Edition; 2015.

Bisciotti GN, Quaglia A, Belli A, et al. Return to sports after ACL reconstruction: a new functional test protocol. Muscles Ligaments Tendons J. 2016;06:499–509.

Bisciotti GN, Volpi P. Return to play. In: Volpi P, editor. Football doctor manual. Trento: Edra Edition; 2018. p. 247–59.

Bojsen-Møller J, Hansen P, Aagaard P, Svantesson U, Kjaer M, Magnusson SP. Differential displacement of the human soleus and medial gastrocnemius aponeurosis during isometric plantar flexor contractions in vivo. J Appl Physiol (1985). 2004 Nov;97(5):1908–14.

Botta G, Piccinetti A, Giontella M, Mancini S. Strengthening of venous pump activity of the sural tricipital in orthopaedics and traumatology by means of a new equipment for vascular exercise. Giornale Italiano di Ortopedia e Traumatologia. 2001;27:84–8.

Burke RE, Levine DN, Salcman M, Tsairis P. Motor units in cat soleus muscle: physiological, histochemical and morphological characteristics "Motor units in cat soleus muscle: physiological, histochemical and morphological characteristics". J Physiol. 1974;238(3):503–14.

Carling C, Le Gall F, Orhant E. A four-season prospective study of muscle strain reoccurrences in a professional football club. Res Sports Med. 2011;19:92–102.

Chamari K, Hachana Y, Ahmed YB, Galy O, Sghaïer F, Chatard JC, Hue O, Wisløff U. Field and laboratory testing in young elite soccer players. Br J Sports Med. 2004 Apr;38(2):191–6.

Chamari K, Moussa-Charai I, Boussaïdi L, Hachana Y, Kauech F, Wisløff U. Appropriate interpretation of aerobic capacity: allometric scaling in adult and young soccer players. Br J Sports Med. 2005 Feb;39(2): 97–101.

Clover J, Wall J. Return-to-play criteria following sports injury. Clin Sports Med. 2010;29:169–75.

Colberg RE, Aune KT, Choi AJ, et al. Incidence and prevalence of musculoskeletal conditions in collegiate tennis athletes. JMST. 2015;20:137–44.

Croisier J-L, Forthomme B, Namurois M-H, et al. Hamstring muscle strain recurrence and strength performance disorders. Am J Sports Med. 2002;30:199–203.

Delvaux F, Rochcongar P, Bruyere O, et al. Return-to-play criteria after hamstring injury: actual medicine practice in professional soccer teams. J Sports Sci Med. 2014;13:721–3.

Glazer DD. Development and preliminary validation of the injury- psychological readiness to return to sport (I-PRRS) scale. J Athl Train. 2009;44:185–9.

Gollnick PD, Sjödin B, Karlsson J, Jansson E, Saltin B. Human soleus muscle: a comparison of fiber composition and enzyme activities with other leg muscles. Pflugers Arch. 1974 Apr 22;348(3):247–55.

Green B, Pizzari T. Calf muscle strain injuries in sport: a systematic review of risk factors for injury. Br J Sports Med. 2017 Aug;51(16):1189–94.

Hachana Y, Chaabene H, Nabli MA, et al. Test-retest reliability, criterion-related validity, and minimal detectable change of the Illinois agility test in male team sport athletes. J Strength Cond Res. 2013;27: 2752–9.

Hagglund M, Walden M, Ekstrand J. Risk factors for lower extremity muscle injury in professional soccer: the UEFA injury study. Am J Sports Med. 2013b;41:327–35.

Hagglund M, Walden M, Magnusson H, et al. Injuries affect team performance negatively in professional football: an 11-year followup of the UEFA champions league injury study. Br J Sports Med. 2013a;47: 738–42.

Hamilton N, Luttgens K. Kinesiology: scientific basis of human motion. 10th ed. New York: McGraw-Hill Higher Education; 2001.

Harris-Love MO, Shrader JA, Davenport TE, et al. Are repeated single-limb heel raises and manual muscle testing associated with peak plantar-flexor force in people with inclusion body myositis? Phys Ther. 2014;94:543–52.

Heck H, Mader A, Hess G, Mucke S, Muller R, Hollmann W. Justification of the 4-mmol/l lactate threshold. Int J Sports Med. 1985;6:117–30.

Hewett T, Snyder-Mackler L, Spindler KP. The drop-jump screening test: difference in lower limb control by gender and effect of neuromuscular training in female athletes. Am J Sports Med. 2007;35:145.

Hoch MC, McKeon PO. Normative range of weight-bearing lunge test performance asymmetry in healthy adults. Man Ther. 2011 Oct;16(5):516–9.

Hodgson JA, Finni T, Lai AM, Edgerton VR, Sinha S. Influence of structure on the tissue dynamics of the human soleus muscle observed in MRI studies during isometric contractions. J Morphol. 2006;267(5): 584–601.

Korkia PK, Tunstall-Pedoe DS, Maffuli N. An epidemiological investigation of training and injury patterns in British triathletes. Br J Sports Med. 1994 Sep;28(3):191–6.

Koulouris G, Ting AYI, Jhamb A, et al. Magnetic resonance imaging findings of injuries to the calf muscle complex. Skelet Radiol. 2007;36:921–7.

Kvist J. Rehabilitation following anterior cruciate ligament injury: current recommendations for sport participation. Sports Med. 2004;34:296–80.

Langford JL, Webster KE, Feller JA. A prospective longitudinal study to assess psychological changes following anterior cruciate ligament reconstruction surgery. Br J Sports Med. 2009;43:377–8.

Luke AC, Kinney SA, D'Hemecourt PA, et al. Determinants of injuries in young dancers. Med Prob Perform Artists. 2002;17(3):105–12.

Lundgren JM, Davis BA. Endartery stenosis of the popliteal artery mimicking gastrocnemius strain: a case report. Arch Phys Med Rehabil. 2004;85(9):1548–51.

Malliaropoulos N, Isinkaye T, Tsitas K, et al. Reinjury after acute posterior thigh muscle injuries in elite track and field athletes. Am J Sports Med. 2011;39:304–10.

Mallo J, González P, Veiga S, Navarro E. Injury incidence in a spanish sub-elite professional football team: a prospective study during four consecutive seasons. J Sports Sci Med. 2011 Dec 1;10(4):731–6.

McCarty EC, Ritchie P, Gill HS, et al. Shoulder instability: return to play. Clin Sports Med. 2004;23:335–51.

Meyers RA, Hermanson JW. Horse soleus muscle: postural sensor or vestigial structure? Anat Rec A Discov Mol Cell Evol Biol. 2006 Oct;288(10):1068–76.

Moller M, Lind K, Styf J, et al. The reliability of isokinetic testing of the ankle joint and a heel-raise test for endurance. Knee Surg Sports Traumatol Arthrosc. 2005;13:60–71.

Negra Y, Chaabene H, Hammami M, et al. Agility in young athletes: is it a different ability from speed and power? J Strength Cond Res. 2017;31:727–35.

Orchard J, Seward H. AFL injury report. Australian Football League (AFL). 2014;2014:1–20.

Orchard JW. Intrinsic and extrinsic risk factors for muscle strains in Australian football. Am J Sports Med. 2001;29:300–3.

Pezzotta G, Querques G, Pecorelli A, Nani R, Sironi S. MRI detection of soleus muscle injuries in professional football players. Skelet Radiol. 2017 Nov;46(11):1513–20.

Powden CJ, Hoch JM, Hoch MC. Reliability and minimal detectable change of the weight-bearing lunge test: a systematic review. Man Ther. 2015 Aug;20(4):524–32.

Powell HC, Silbernagel KG, Brorsson A, et al. Individuals post Achilles tendon rupture exhibit asymmetrical knee and ankle kinetics and loading rates during a drop countermovement jump. J Orthop Sports Phys Ther. 2018;48:34–43.

Raya MA, Gailey RS, Gaunaurd IA, et al. Comparison of three agility tests with male servicemembers: Edgren side step test, t-test, and Illinois agility test. J Rehabil Res Dev. 2013;50:951–60.

Rebella G. A prospective study of injury patterns in collegiate pole vaulters. Am J Sports Med. 2015;43: 808–15.

Reurink G, Goudswaard GJ, Tol JL, et al. MRI observations at return to play of clinically recovered hamstring injuries. Br J Sports Med. 2014;48(18):1370–6.

Saladin KS. Anatomy & physiology: the unity of form and function. 6th ed. New York: McGraw-Hill Higher Education; 2007.

Sanfilippo JL, Silder A, Sherry MA, et al. Hamstring strength and morphology progression after return to sport from injury. Med Sci Sports Exerc. 2013;45: 448–54.

Sassi RH, Dardouri W, Yahmed MH, Gmada N, Mahfoudhi ME, Gharbi Z. Relative and absolute reliability of a modified agility T-test and its relationship with vertical jump and straight sprint. J Strength Cond Res. 2009 Sep;23(6):1644–51.

Silbernagel KG, Gustavsson A, Thomee R, et al. Evaluation of lower leg function in patients with Achilles tendinopathy. Knee Surg Sports Traumatol Arthrosc. 2006;14:1207–17.

Witvrouw E, Danneels L, Asselman P, et al. Muscle flexibility as a risk factor for developing muscle injuries in male professional soccer players. Am J Sports Med. 2003;31:41–6.

Zambaldi M, Beasley I, Rushton A. Return to play criteria after hamstring muscle injury in professional football: a Delphi consensus study. Br J Sports Med. 2017;51:1221–6.

12 Return to Training and Return to Play Following Hip Short External Rotator Muscle Injury

12.1 Anatomical Description

Six muscles situated deeply on the dorsal side of the coxofemoral joint compose the hip short external rotator muscles. Considering their insertion, it is possible to divide the hip short external rotator muscles in two groups (Bouchet 1996; Rouvière 2002):

The first group, with muscles that originate from the pelvis, is composed of:

1. The piriformis.
2. The internal obturator.

The second group, with muscles that originate outside of the pelvis, is composed of:

1. The gemellus muscles (superior and inferior).
2. The quadratus femoris.
3. The external obturator.

12.1.1 The Piriformis

The piriformis (derived from Latin *piriformis*, meaning "pear-shaped") was for the first time described by the Flemish anatomist Adrianus Spigelius (1578–1625), during his period of work At Padua University (Italy) under the guidance of the famous Italian anatomist Girolamo Fabrizi d'Acquapendente (1533–1619).

The piriformis is a flat muscle, pyramidal in shape, lying almost parallel with the posterior margin of the gluteus medius. The muscle originates, by three fleshy digitations, from the anterior part of the sacrum in the anatomical region located laterally to second, third, and fourth ventral sacral foramina (S2-S4) and from the edge of the greater ischial. Furthermore, a few fibers also arise from the anterior surface of the sacrotuberous ligament.

The muscle passes out of the pelvis through the greater sciatic foramen and, travelling transversally and dorsally to the hip joint, inserts distally on the greater trochanter of the femur. The piriformis tendon often joins with the tendons of the superior gemellus, inferior gemellus, and internal obturator muscles prior to the insertion. In 17% of people, the piriformis muscle is pierced by the sciatic nerve. Several variations may occur in its anatomy, but the most common type of anomaly (81% of anomalies) is the Beaton's type B, in which the common peroneal nerve pierces the piriformis muscle (Smoll 2010). Some other types of anatomical variants provide that it may be united with the gluteus medius, send fibers to the gluteus minimus, or receive fibers from the superior gemellus (Smoll 2010). The piriformis represents an important anatomical landmark in the gluteal region. As it travels through the greater sciatic foramen, it divides it into an inferior and superior part. Indeed, it determines the name of

G. N. Bisciotti et al., *Return-to-Play after Lower Limb Muscle Injury in Football*,
https://doi.org/10.1007/978-3-030-84950-4_12

the vessels and nerves in this region: the nerve and vessels emerging superiorly to the piriformis are the superior gluteal nerve and the superior gluteal vessels. On the contrary, the nerve and vessels emerging inferiorly to the piriformis are the inferior gluteal nerve and the inferior gluteal vessels.

The piriformis externally rotates the femur with hip extension and abducts the femur with hip flexion. The abduction of the flexed thigh is an important action during walking because it shifts the body weight to the opposite side when the foot is lifted, preventing falling. Furthermore, still during walking, the piriformis synchronizes the movement of the sacrum in relationship with the iliac bone, thus preventing overuse of the sacroiliac joints (Delin et al. 2017). The piriformis is innervated by branches of the sacral plexus (first and second sacral nerves), and its blood supply comes from the superior gluteal artery.

12.1.2 The Internal Obturator

The internal obturator originates on the hip internal surface (rim of the obturator foramen and internal surface of the obturator membrane). Its shape is triangular with a base oriented medially and ventrally. It exits the pelvic cavity through the lesser sciatic foramen ending in four or five tendinous bands. Distally it inserts on the greater trochanter of the femur at the level of the trochanteric fossa. It is innervated by the nerve of the internal obturator muscle, a collateral branch of the sacral plexus (S1 and S2). Its blood supply comes from the obturator artery. The internal obturator is an external rotator and accessory abductor of the thigh in a flexed position at the level of the hip joint.

12.1.3 Gemellus Muscles

The gemellus are two muscles (superior gemellus and inferior gemellus) surrounding the terminal portion of the internal obturator muscle. The etymology of the name comes from the Latin word "*geminus*" meaning twin, doubled, or duplicated. Indeed, the superior and inferior gemellus are paired muscles performing the same action.

12.1.3.1 Superior Gemellus

The superior gemellus, the smaller of the two, arises from the external (gluteal) surface of the spine of the ischium blending with the upper part of the tendon of the internal obturator muscle. Distally it finishes on the terminal tendon of the internal obturator muscle, not having its own terminal insertion. In an anatomical variant, the superior gemellus fibers may extend further than average, prolonging onto the medial surface of the greater trochanter of the femur. The superior gemellus is innervated by the nerve of the internal obturator muscle. The blood supply comes from the lower gluteal artery.

12.1.3.2 Inferior Gemellus

The inferior gemellus arises from the upper part of the tuberosity of the ischium immediately below the groove for the internal obturator tendon. It blends with the lower part of the tendon of the internal obturator. Like the gemellus superior, it distally inserts on the terminal tendon of the internal obturator muscle, not having its own terminal insertion. The inferior gemellus is innervated by the nerve of the quadratus femoris muscle. The blood supply comes from the lower gluteal artery.

The gemellus superior and gemellus inferior help to laterally rotate the extended thigh. Both muscles also help to steady the femoral head in the acetabulum by assisting the internal obturator muscle action (Palastanga and Soames 2011). Indeed, as the internal obturator turning around the lesser sciatic notch loses some of its power, this is compensated for by the action of the gemelli (Palastanga and Soames 2011).

12.1.4 The Quadratus Femoris

The quadratus femoris is a flat, quadrilateral muscle located on the posterior side of the hip joint. It originates on the lateral border of the ischial tuberosity. At its origin, it is separated from the upper margin of the adductor by the terminal branches of the medial femoral circumflex vessels. Distally, it inserts on the quadrate tubercle of the intertrochanteric crest. A bursa is often found between the front of quadratus femoris and the lesser trochanter. It is innervated by the nerve of the quadratus femoris (collateral of the sacral plexus, S1and S2). The quadratus femoris is an external rotator and accessory adductor.

12.1.5 The External Obturator

The external obturator muscle is a flat, triangular muscle, which covers the outer surface of the anterior wall of the pelvis. It arises from the lateral surface of the hip bone, immediately around the external surface of the obturator membrane and from the surrounding bone. The fibers are medially oriented, passing first obliquely by the inferior and subsequently by the posterior surfaces of the femoral neck and then dorsally in the external obturator groove. Distally, it inserts into the trochanteric fossa of the femur. The muscle is innervated by the obturator nerve. Its blood supply comes from the obturator artery. Its action externally rotates the hip. Furthermore, it is an accessory adductor.

12.2 Epidemiological Notes

Despite the important role of the hip short external rotator muscle group, little is found in literature. The existing studies mainly describe chronic syndromes such as piriformis syndrome or ischial-femoral impingement (Meknas et al. 2009), and only a few studies are based on acute injuries. However, the main mechanisms causing traumatic injury to this muscles group are essentially three:

1. The first is an unstable pelvis position with a sudden change in body weight distribution (Cass 2015; Wong-On et al. 2018).
2. The second is a sudden change of direction in condition of pelvic instability (Cass 2015; Wong-On et al. 2018).
3. The third is represented by a sudden hip intrarotation with hip and knee flexed at about 90° (Delp et al. 1999).

The hip short external rotator injuries are frequently associated with other muscle injuries (i.e., rectus femoris, iliopsoas, great adductor, and hamstrings). Furthermore, concomitant injuries to several of the hip short external rotator muscles are possible but infrequent (Delin et al. 2017).

Reinjuries of the hip short external rotator muscles not being described in literature (Wong-On et al. 2018) confirm the hypothesis that such reinjuries are dependent on a well-defined mechanical situation, like those described above, and are independent of predisposing intrinsic factors, such as excessive retraction or stiffness of the considered muscles (Wong-On et al. 2018). The hip short external rotator muscle injuries present a good prognosis and a relatively short recovery time (Byrne et al. 2017; Wong-On et al. 2018).

12.3 Clinical and Imaging Assessments for RTT

The tests suggested by the CC following short external hip rotator muscles injuries are:

12.3.1 General Assessment

Below are the Consensus Conference recommendations for clinical and imaging assessments for RTT following short external hip rotator muscles injuries:

1. Absence of clinical symptoms (Kvist 2004; Malliaropoulos et al. 2011; Delvaux et al. 2014).
2. Absence of pain or tenderness during muscle palpation (Kvist 2004; Zambaldi et al. 2017; Bisciotti and Volpi 2018; Reurink et al. 2014).
3. Absence of pain on passive and active stretching (Witvrouw et al. 2003; Bisciotti 2015; Bisciotti and Volpi 2018).

4. Absence of pain on isometric, concentric, and eccentric contraction (Bisciotti and Volpi 2018).
5. Completion of the prescribed rehabilitation program (Reurink et al. 2014).
6. MRI and US imaging assessment points will be specified in Chap. 5: "The Role of Imaging in the Return to Training and Return to Play Decision-Making Process."
7. Subjective feelings of the player taken into account (i.e., assess levels of anxiety, apprehension, fear of failure, and/or fear of reinjury) (McCarty et al. 2004; Bauman 2005; Glazer 2009; Langford et al. 2009; Clover and Wall 2010; Delvaux et al. 2014).

12.3.2 Specific Assessment

The specific assessment for short external hip rotator muscles injury is based on the following tests (Delp et al. 1999; Busfield and Romero 2009; Valente et al. 2011; Delin et al. 2017):

1. Beatty maneuver test.
2. Freiberg maneuver test.
3. Internal rotation test.

12.4 Laboratory Tests for RTT

The following are the recommended laboratory tests for RTT after short external hip rotator muscles injury:

1. Pace and Nagle maneuvers assessed by dynamometric tests (Delp et al. 1999; Croisier et al. 2002; Sanfilippo et al. 2013; Delvaux et al. 2014; Bisciotti et al. 2016).

12.5 Field Tests for RTT

The following are the field tests prior to RTT after hip short external rotator muscle injuries recommended by the Italian consensus statement on return to play after lower limb muscle injury in football (Bisciotti et al. 2019),

1. Kicking test (Bisciotti et al. 2016).
2. Carioca test (Kong et al. 2012; Jang et al. 2014).
3. Illinois Agility test (Hachana et al. 2013; Raya et al. 2013; Bisciotti 2015; Negra et al. 2017; Bisciotti et al. 2018).
4. Agility T-test (Sassi et al. 2009).

12.6 RTP Tests After Short External Hip Rotator Muscles Injuries

As in the case of all the other muscles previously considered, the criteria for RTP-DMP are related to the evaluation of the football performance. Therefore the RTP-DMP will chronologically follow the RTT-DMP only if the athlete has obtained a positive judgment for the RTT. In such case, the period of registration of the tests for RTP begins.

In this regard, the CC recommend the following points:

1. The data acquisition period must start from the first day of RTP and last 7–10 days.
2. During this period, the performance data must be systematically recorded via the GPS system.
3. It is necessary to identify some "typical" sessions, substantially overlapping each other, which can be deduced both from the last prelesional week and from the post-period.

As already mentioned in Chap. 4, the three evaluation categories are:

1. Quantitative evaluation.
2. Qualitative evaluation.
3. Parameter analysis.

All the indices described in Chap. 4 as related to the abovementioned categories must be analyzed. The reference value below which the positive judgment for RTP is postponed is set at a maximum difference of 10% between the prelesional data and the data recorded during the acquisition period following the RTT. For SGMC injuries too, concerning the control of aerobic fitness, the CC experts suggest that the RTP-DMP be implemented by means of a test for VO_2max determination (Heck et al. 1985; Chamari et al. 2004; Chamari et al. 2005; Badawy and Muaidi 2018), in which the player must record a value of VO_2max, and consequently of VAM, equal to at least 90% of the pre-injury values.

References

Bauman J. Returning to play: the mind does matter. Clin J Sport Med. 2005;15:432–5.

Bisciotti GN, Volpi P, Amato M, Alberti G, Allegra F, Aprato A, et al. Italian consensus conference on guidelines for conservative treatment on lower limb muscle injuries in athlete. BMJ Open Sport Exerc Med. 2018 May 24;4(1):e000323.

Bisciotti GN, Volpi P, Zini R, et al. Groin pain syndrome Italian consensus conference on terminology, clinical evaluation and imaging assessment in groin pain in athlete. BMJ Open Sport Exerc Med. 2016;2(1):e000142.

Bisciotti GN, Volpi P. Return to play. In: Volpi P, editor. Football doctor manual. Trento: Edra Edition; 2018. p. 247–59.

Bisciotti GN. Return to play after a muscle lesion. In: Volpi P, editor. Arthroscopy in sport. Cham: In, Springer Edition; 2015.

Bisciotti GN, Volpi P, Alberti G, Aprato A, Artina M, Auci, et al. Italian consensus statement (2020) on return to play after lower limb muscle injury in football (soccer). BMJ Open Sport Exerc Med. 2019 Oct 15;5(1):e000505.

Bouchet A. Anatomie topographique descriptive et fonctionnelle, vol. 3(b). 3rd ed. Paris: SIMEP; 1996.

Busfield BT, Romero DM. Obturator internus strain in the hip of an adolescent athlete. Am J Orthop (Belle Mead NJ). 2009 Nov;38(11):588–9.

Byrne C, Alkhayat A, O'Neill P, Eustace S, Kavanagh E. Obturator internus muscle strains. Radiol Case Rep. 2017 Mar;12(1):130–2.

Cass SP. Piriformis syndrome: a cause of non-discogenic sciatica. Curr Sports Med Rep. 2015 Jan;14(1):41.

Chamari K, Hachana Y, Ahmed YB, Galy O, Sghaïer F, Chatard JC, Hue O, Wisløff U. Field and laboratory testing in young elite soccer players. Br J Sports Med. 2004 Apr;38(2):191–6.

Chamari K, Moussa-Charai I, Boussaïdi L, Hachana Y, Kauech F, Wisløff U. Appropriate interpretation of aerobic capacity: allometric scaling in adult and young soccer players. Br J Sports Med. 2005 Feb;39(2):97–101.

Clover J, Wall J. Return-to-play criteria following sports injury. Clin Sports Med. 2010;29:169–75.

Croisier J-L, Forthomme B, Namurois M-H, et al. Hamstring muscle strain recurrence and strength performance disorders. Am J Sports Med. 2002;30:199–203.

Delin C, Vandensteen JY, Roger B. Hip short external rotator muscles injuries. In: Roger B, Guermazi A, Skaf A, editors. Muscle injuries in sport athletes. Cham: Springer Editions; 2017.

Delp SL, Hess WE, Hungerford DS, Jones LC. Variation of rotation moment arms with hip flexion. J Biomech. 1999;32(5):493–501.

Delvaux F, Rochcongar P, Bruyere O, et al. Return-to-play criteria after hamstring injury: actual medicine practice in professional soccer teams. J Sports Sci Med. 2014;13:721–3.

Glazer DD. Development and preliminary validation of the injury- psychological readiness to return to sport (I-PRRS) scale. J Athl Train. 2009;44:185–9.

Hachana Y, Chaabene H, Nabli MA, et al. Test-retest reliability, criterion-related validity, and minimal detectable change of the Illinois agility test in male team sport athletes. J Strength Cond Res. 2013;27:2752–9.

Heck H, Mader A, Hess G, Mucke S, Muller R, Hollmann W. Justification of the 4-mmol/l lactate threshold. Int J Sports Med. 1985;6:117–30. J Orthop. 2009;38(11):588–9

Jang SH, Kim JG, Ha JK, et al. Functional performance tests as indicators of returning to sports after anterior cruciate ligament reconstruction. Knee. 2014;21: 95–101.

Kong DH, Yang SJ, Ha JK, et al. Validation of functional performance tests after anterior cruciate ligament reconstruction. Knee Surg Relat Res. 2012;24:40–5.

Kvist J. Rehabilitation following anterior cruciate ligament injury: current recommendations for sport participation. Sports Med. 2004;34:296–80.

Langford JL, Webster KE, Feller JA. A prospective longitudinal study to assess psychological changes following anterior cruciate ligament reconstruction surgery. Br J Sports Med. 2009;43:377–8.

Badawy MM, Muaidi QI. Aerobic profile during high-intensity performance in professional Saudi athletes. Pak J Biol Sci. 2018;21(1):24–8.

Malliaropoulos N, Isinkaye T, Tsitas K, et al. Reinjury after acute posterior thigh muscle injuries in elite track and field athletes. Am J Sports Med. 2011;39(2): 304–10.

McCarty EC, Ritchie P, Gill HS, et al. Shoulder instability: return to play. Clin Sports Med. 2004;23: 335–51.

Meknas K, Kartus J, Letto J, Christensen A, Johansen O. Surgical release of the internal obturator tendon for the treatment of retro-trochanteric pain syndrome: a prospective randomized study, with long-term follow-up. Knee Surg Sports Traumatol Arthrosc. 2009;17:1249–56.

Negra Y, Chaabene H, Hammami M, et al. Agility in young athletes: is it a different ability from speed and power? J Strength Cond Res. 2017;31:727–35.

Palastanga N, Soames R. Physiotherapy essentials: anatomy and human movement: structure and function. 6th ed. London: GBR: Elsevier Health Sciences; 2011.

Raya MA, Gailey RS, Gaunaurd IA, et al. Comparison of three agility tests with male servicemembers: Edgren side step test, t-test, and Illinois agility test. J Rehabil Res Dev. 2013;50:951–60.

Reurink G, Goudswaard GJ, Tol JL, et al. MRI observations at return to play of clinically recovered hamstring injuries. Br J Sports Med. 2014;48(18):1370–6.

Rouvière H. L'Anatomie Humaine, vol. 3. Paris: Elsevier-Masson; 2002.

Sanfilippo JL, Silder A, Sherry MA, et al. Hamstring strength and morphology progression after return to sport from injury. Med Sci Sports Exerc. 2013;45:448–54.

Sassi RH, Dardouri W, Yahmed MH, Gmada N, Mahfoudhi ME, Gharbi Z. Relative and absolute reliability of a modified agility T-test and its relationship with vertical jump and straight sprint. J Strength Cond Res. 2009 Sep;23(6):1644–51.

Smoll NR. Variations of the piriformis and sciatic nerve with clinical consequence: a review. Clin Anat. 2010 Jan;23(1):8–17.

Valente HC, Marques FO, Da Silva De Souza L, Abib RT, Ribeiro DC. Injury of the external obturator muscle in professional soccer athletes. Rev Bras Med Esporte. 2011;17(1):36–9.

Witvrouw E, Danneels L, Asselman P, et al. Muscle flexibility as a risk factor for developing muscle injuries in male professional soccer players. Am J Sports Med. 2003;31:41–6.

Wong-On M, Turmo-Garuz A, Arriaza R, Gonzalez de Suso JM, Til-Perez L, Yanguas-Leite X, Diaz-Cueli D, Gasol-Santa X. Injuries of the obturator muscles in professional soccer players. Knee Surg Sports Traumatol Arthrosc. 2018 Jul;26(7):1936–42.

Zambaldi M, Beasley I, Rushton A. Return to play criteria after hamstring muscle injury in professional football: a Delphi consensus study. Br J Sports Med. 2017;51:1221–6.

13 Return to Training and Return to Play Following Iliopsoas Injury

13.1 Anatomical Description

The iliopsoas tendon-muscle is a complex composed of four muscles: the iliacus, ilioinfratrochanteric, the psoas major, and the psoas minor.

The iliacus is a triangular fan-shaped muscle that arises from the iliac fossa and from the region of the anterior inferior iliac spine. The muscle is innervated by the femoral nerve (which is composed of nerves from the L2-L4 anterior rami). Another smaller bundle, called the ilioinfratrochanteric muscle, has been observed by some authors lateral to the lateral fibers of the iliacus muscle (Tatu et al. 2001; Guillin et al. 2008). The iliacus muscle blood supply comes from the arteria iliolumbalis, the arteria obturatoria, the arteria iliaca externa, and the arteria femoralis (Moore and Dalley 1999; Philippon et al. 2014a, b).

The psoas major is a long muscle, fusiform in shape, originating from T12, LS, and L6 (Moore and Dalley 1999; Philippon et al. 2014a, b). The muscle is innervated by direct branches of the anterior rami of the lumbar plexus at L1-L3 level. The blood supply comes from the arteria iliolumbalis, the arteria obturatoria, the arteria iliaca externa, and the arteria femoralis.

The psoas minor originates from the vertical fascicles inserted on the last thoracic and first lumbar vertebrae. Then, the muscle passes down onto the medial border of the psoas major and is inserted to the innominate line and the iliopectineal eminence. Additionally, the psoas minor attaches to and stretches the deep surface of the iliac fascia. Furthermore, occasionally its lowermost fibers reach the inguinal ligament. The muscle is supplied by the four lumbar arteries and by the lumbar branch of the iliolumbar artery. The psoas minor is a weak flexor of the lumbar vertebral column. The psoas minor is considered inconstant and is often absent, being present in only 40%–65% of the human specimens studied (Clemente 1997; Farias et al. 2012; Neumann and Garceau 2014).

The psoas major and the iliacus muscles converge at the level of L5-S2 forming the iliopsoas muscle. The psoas major and the iliacus muscles insert on the lesser trochanter of the femur. The inferior portion of the iliopsoas muscle below the inguinal ligament forms part of the floor of the femoral triangle. The iliopsoas muscle is covered by the iliac fascia, which surround the psoas major muscle as it passes under the medial arcuate ligament. Together with the iliac fascia, it continues down to the inguinal ligament forming the iliopectineal arch, which separates the muscular and vascular lacunae. The iliopsoas is a typical posture muscle composed mostly of slow-twitch fibers (Moore and Dalley 1999; Philippon et al. 2014a, b). Some interesting anatomical peculiarity of the iliopsoas are the femoral nerve passes through the iliopsoas and innervates the quadriceps, the pectineus, and the sartorius muscles; the obturator nerve too passes through the iliopsoas and is responsible for the sensory innervation

G. N. Bisciotti et al., *Return-to-Play after Lower Limb Muscle Injury in Football*,
https://doi.org/10.1007/978-3-030-84950-4_13

of the skin of the medial aspect of the thigh and the motor innervation of the adductor muscles of the lower extremity (i.e., the external obturator, the adductor longus, the adductor magnus, the adductor brevis, and the gracilis muscles). Any of these nerve structures during their course within the iliopsoas might be affected. The iliopsoas also shows some interesting anatomical variants.

Philippon et al. (2014a, b) in a study based on the examination of fresh frozen cadavers demonstrated the existence at the level of the hip joint of a single-, double-, and triple-banded iliopsoas tendon in 28.3%, 64.2%, and 7.5% of the cases, respectively, while Crompton et al. (2014) showed the presence of two distinct tendons in 21% of the patients belonging to pediatric population. The iliopsoas bursa (also called iliopectineal bursa) is positioned between the musculotendinous unit and the bony surfaces of proximal femur and the pelvis. The iliopectineal bursa is the largest bursa of the human body, extending from the iliopectineal eminence to the lower portion of the femoral head. It is on average from 5 to 6 cm long and 3 cm wide (Tatu et al. 2001). The possibility of a communication of the iliopectineal bursa with the hip joint through a congenital defect between the iliofemoral and pubofemoral ligaments is controversy. Tatu et al. (2001) in a study based on 14 cadaveric dissections did not find any communication. On the contrary, Schaberg et al. (1984) observed a direct communication between the joint and the bursa in 15% of the patients considered in their study. The presence of any communication between the bursa and the hip joint is an important point to consider during the diagnostic injection. Indeed, if the anesthetic can seep through the intra-articular and bursal compartments, it could confuse the diagnostic result.

From a functional point of view, the iliopsoas is the prime mover of hip flexion and is the strongest hip flexor playing a vital role in the biomechanics of standing, walking, and running. The muscle has also an important role during the femoral external rotation end in the lateral bending, flexion, and during the balance of the upper body (Fitzgerald 1969; Mann et al. 1986; Rajendran 1989; Andersson et al. 1995). Some studies show that the iliacus and psoas major have different individual and task-specific activation patterns. In particular, the iliacus muscle is an important pelvis stabilizer and plays a crucial role during the rapid hip flexion movements while running (Serner et al. 2015). The psoas major plays an important role for sitting in an erect position and for the stability of the spine in the frontal plane (Andersson et al. 1995). During the sit-up movement, the contribution of the iliacus and psoas muscles depends upon the angle of hip flexion (Andersson et al. 1995).

13.2 Epidemiological Notes

The most frequent injury mechanism in soccer is the kicking (40% of all football injuries occur during kicking), while the change of direction is the most frequent in other sports (31%). During this type of movements, adductor injuries accounted for 66% of all injuries and primarily involved the adductor longus. In the same movements, also the iliopsoas muscle is frequently injured according to both MRI and US examination (Anderson 2016). Some studies report that the iliopsoas is the primary cause of acute groin pain syndrome in 25%–30% of the cases (Hölmich and Renstrom 2007; Hölmich et al. 2014; Rankin et al. 2015). Iliopsoas muscle injuries are single in the majority of cases but can also be coupled to adductor longus (15.9%), rectus femoris (8.5%), rectus abdominis (2.4%), and sartorius (2.4%) injuries (Serner et al. 2015). From a radiological point of view, iliopsoas injuries can be divided into injuries in the iliacus muscle, in the psoas muscle or in both of them. The injuries involving the iliacus muscle are the most frequent. The iliopsoas muscle is generally considered to form a common tendon at its distal insertion. Yet, recent studies have shown the existence of a separate tendon that is present in the majority of cases. Furthermore, an accessory iliacus tendon can sometimes be present (Tatu et al. 2001; Polster et al. 2007; Philippon et al. 2014a, b). The diagnostic importance and the clinical relevance of that remains unknown. Internal hip snapping (also known as coxa saltans interna) is a clinical situation that deserves a separate mention. The internal hip snapping was for the first time described by Nunziata and Blumenfeld in 1951 (Nunziata and Blumenfeld 1951). The authors described the phenomenon as an audible snapping of the iliopsoas tendon over the iliopectineal eminence. The snapping is audible during the following movement: the patient from a position of flexion, abduction, and external rotation brings the leg into extension and neutral position. However, proposals of alternative mechanisms to explain internal hip snapping can be found in the literature. Some authors talk about an accessory iliopsoas tendinous slip (Deslandes et al. 2008) or the presence of a paralabral cyst (Deslandes et al. 2008), while other studies described a situation of stenosing tenosynovitis (Micheli 1983). Furthermore, other studies have observed that an internal hip snapping occurs over the lesser trochanter (Schaberg et al. 1984) and the femoral head (Howse 1972). Some relatively more recent studies based on dynamic US showed, as cause of snapping, a sudden flipping of the psoas tendon over the iliacus muscle (Winston et al. 2006; Deslandes et al. 2008; Guillin et al. 2008). In any case, at today, the exact internal hip snapping mechanism is still controversial and lacks consensus. Despite the lack of a precise etiopathogenesis of the phenomenon, iliopsoas bursitis (Fig. 13.1) and tendinopathy are closely associated to internal hip snapping (Schaberg et al. 1984; Johnston et al. 1998; Pelsser et al. 2001; Wunderbaldinger et al. 2001). These situations are so strongly associated that some authors defined this clinical condition as "iliopsoas syndrome" (Johnston et al. 1998), even though others studies did not show abnormality in patients undergoing surgery for internal symptomatic hip snapping (Schaberg et al. 1984; Jacobson and Allen 1990). Moreover, the internal hip snapping is associated to an acute injury in 50% of the cases. Finally, in general population, the prevalence of an asymptomatic internal hip snapping is on average equal to 40% (Guillin et al. 2008). The last remarkable clinical situation regarding iliopsoas is the iliopsoas impingement. It was first described by Heyworth et al. (2007) and is a mechanical situation in which an excessively tight iliopsoas tendon causes a mechanical impingement with the underling acetabular labrum. This results in a labral abnormality at the level of the iliopsoas tendon passage over the acetabular rim.

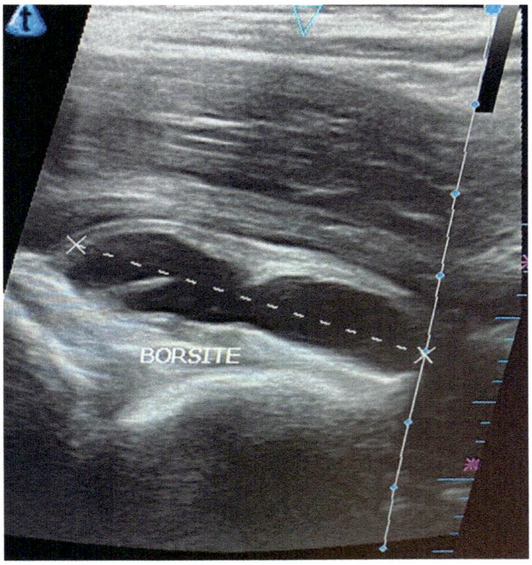

Fig. 13.1 US image of iliopsoas bursitis

13.3 Clinical and Imaging Assessments for RTT

The tests suggested by the CC following iliopsoas indirect injury are:

13.3.1 General Assessment

1. Absence of clinical symptoms (Kvist 2004; Malliaropoulos et al. 2011; Delvaux et al. 2014).
2. Absence of pain or tenderness during muscle palpation (Kvist 2004; Zambaldi et al. 2017; Bisciotti and Volpi 2018; Reurink et al. 2014).
3. Absence of pain on passive and active stretching (Witvrouw et al. 2003; Bisciotti and Volpi 2018).
4. Absence of pain on isometric, concentric, and eccentric contraction (Bisciotti and Volpi 2018).
5. Completion of the prescribed rehabilitation program (Reurink et al. 2014).

6. MRI and US imaging assessment points will be specified in Chap. 5: "The Role of Imaging in the Return to Training and Return to Play Decision-Making Process."
7. Subjective feelings of the player taken into account (i.e., assess levels of anxiety, apprehension, fear of failure, and/or fear of reinjury) (McCarty et al. 2004; Bauman 2005; Glazer 2009; Langford et al. 2009; Clover and Wall 2010; Delvaux et al. 2014).

13.3.2 Specific Assessment

1. Thomas test (Harvey 1998; Ferber and Kendall 2010).

13.4 Laboratory Tests for RTT

After iliopsoas muscle injury, the following are the laboratory tests for RTT recommended by CC:

1. Iliopsoas muscle strength assessed by dynamometric tests. The values of dynamometric tests must be >90% of pre-lesional or contralateral values.

13.5 Field Tests for RTT

The following field tests are recommended for RTT-DMP after iliopsoas muscle injury:

1. Illinois Agility Test (Hachana et al. 2013; Raya et al. 2013; Bisciotti and Volpi 2018).
2. Kicking test (Bisciotti and Volpi 2018).

13.6 RTP Tests after Iliopsoas Injury

As for the other muscle injuries described, the criterion for RTP-DMP is performance evaluation. The RTP-DMP chronologically follows the RTT-DMP. Therefore, the athlete can obtain clearance for RTP only after the RTT-DMP. With the RTP-DMP the period of GPS registration begins.

In this regard, the CC recommends, as for the other muscle injuries described, the following requirements:

1. The data acquisition period must start from the first day of RTP and include a period of 7–10 days.
2. During this period, the performance data must be systematically recorded via the GPS system.
3. Identification of some "typical" sessions, substantially overlapping each other, which can be deduced from the pre-lesional period.

As already mentioned in Chap. 4, the three evaluation categories are:

1. Quantitative evaluation.
2. Qualitative evaluation.
3. Parameter analysis.

All indices belonging to the three mentioned categories must be carefully analyzed.

The maximal difference between the prelesional data and the data recorded during the acquisition period following the RTT is fixed in ≤10%.

If the absence from team training was longer than 20 days, the CC experts suggest that the RTP-DMP be implemented by means of a test for VO_2max determination (Chamari et al. 2004; Chamari et al. 2005; Badawy and Muaidi 2018), in which the player must record a value of VO_2max, and consequently of VAM, equal to at least 90% of the pre-injury values.

References

Anderson CN. Iliopsoas: pathology, diagnosis, and treatment. Clin Sports Med. 2016 Jul;35(3):419–33.

Andersson E, Oddsson L, Grundstro MH, et al. The role of the psoas and iliacus muscles for stability and movement of the lumbar spine, pelvis and hip. Scand J Med Sci Sports. 1995;5(1):10–6.

Bauman J. Returning to play: the mind does matter. Clin J Sport Med. 2005;15:432–5.

Bisciotti GN, Volpi P. Return to play. In: Volpi P, editor. Football doctor manual. Trento: Edra Edition; 2018. p. 247–59.

Chamari K, Hachana Y, Ahmed YB, Galy O, Sghaïer F, Chatard JC, Hue O, Wisløff U. Field and laboratory testing in young elite soccer players. Br J Sports Med. 2004 Apr;38(2):191–6.

Chamari K, Moussa-Charai I, Boussaïdi L, Hachana Y, Kauech F, Wisløff U. Appropriate interpretation of aerobic capacity: allometric scaling in adult and young soccer players. Br J Sports Med. 2005 Feb;39(2):97–101.

Clemente CD. Anatomy: a regional atlas of the human body. 4th ed. Baltimore: Lippincott Williams & Wilkins; 1997. p. 363.

Clover J, Wall J. Return-to-play criteria following sports injury. Clin Sports Med. 2010;29:169–75.

Crompton T, Lloyd C, Kokkinakis M, et al. The prevalence of bifid iliopsoas tendon on MRI in children. J Child Orthop. 2014;8(4):333–6.

Delvaux F, Rochcongar P, Bruyère O, Bourlet G, Daniel C, Diverse P, Reginster JY, Croisier JL. Return-to-play criteria after hamstring injury: actual medicine practice in professional soccer teams. J Sports Sci Med. 2014 Sep 1;13(3):721–3.

Deslandes M, Guillin R, Cardinal E. The snapping iliopsoas tendon: new mechanisms using dynamic sonography. AJR Am J Roentgenol. 2008;190(3):576–81.

Farias MC, Riveiro De Oliveira DR, Rocha TD, Caiaffo Brito TD. Morphological and morphometric analysis of psoas minor muscle in cadavers. J Morphol Sci. 2012;29(4):202–5.

Ferber R, Kendall KD, Mc Elroy LJ. Normative and critical criteria for iliotibial band and iliopsoas muscle flexibility. Athl Train. 2010 Jul-Aug;45(4):344–8.

Fitzgerald P. The action of the iliopsoas muscle. Ir J Med Sci. 1969;8(1):31–3.

Glazer DD. Development and preliminary validation of the injury- psychological readiness to return to sport (I-PRRS) scale. J Athl Train. 2009;44:185–9.

Guillin R, Cardinal E, Bureau NJ. Sonographic anatomy and dynamic study of the normal iliopsoas musculotendinous junction. Eur Radiol. 2008;19(4):995–1001.

Hachana Y, Chaabene H, Nabli MA, et al. Test-retest reliability, criterion-related validity, and minimal detectable change of the Illinois agility test in male team sport athletes. J Strength Cond Res. 2013;27:2752–9.

Heyworth BE, Shindle MK, Voos JE, et al. Radiologic and intraoperative findings in revision hip arthroscopy. Arthroscopy. 2007;23(12):1295–302.

Harvey D. Assessment of the flexibility of elite athletes using the modified Thomas test. Br J Sports Med. 1998 Mar;32(1):68–70.

Hölmich P, Renstrom PA. Long-standing groin pain in sportspeople falls into three primary patterns, a "clinical entity" approach: a prospective study of 207 patients. Br J Sports Med. 2007;41(4):247–52.

Hölmich P, Thorborg K, Dehlendorff C, et al. Incidence and clinical presentation of groin injuries in sub-elite male soccer. Br J Sports Med. 2014;48(16):1245–50.

Howse AJ. Orthopaedists aid ballet. Clin Orthop Relat Res. 1972;89:52–63.

Jacobson T, Allen WC. Surgical correction of the snapping iliopsoas tendon. Am J Sports Med. 1990;18(5):470–4.

Johnston CA, Wiley JP, Lindsay DM, et al. Iliopsoas bursitis and tendinitis. A Rev Sports Med. 1998;25(4):271–83.

Kvist J. Rehabilitation following anterior cruciate ligament injury: current recommendations for sport participation. Sports Med. 2004;34:296–80.

Langford JL, Webster KE, Feller JA. A prospective longitudinal study to assess psychological changes following anterior cruciate ligament reconstruction surgery. Br J Sports Med. 2009;43:377–8.

Badawy MM, Muaidi QI. Aerobic profile during high-intensity performance in professional saudi athletes. Pak J Biol Sci. 2018;21(1):24–8.

Malliaropoulos N, Isinkaye T, Tsitas K, et al. Reinjury after acute posterior thigh muscle injuries in elite track and field athletes. Am J Sports Med. 2011;39(2):304–10.

Mann RA, Moran GT, Dougherty SE. Comparative electromyography of the lower extremity in jogging, running, and sprinting. Am J Sports Med. 1986;14(6):501–10.

McCarty EC, Ritchie P, Gill HS, et al. Shoulder instability: return to play. Clin Sports Med. 2004;23:335–51.

Micheli LJ. Overuse injuries in children's sports: the growth factor. Orthop Clin North Am. 1983;14(2):337–60.

Moore KL, Dalley AF. Clinically oriented anatomy. 4th edition. Baltimore: Lippincott Williams & Wilkins; 1999. p. 533.

Neumann DA, Garceau LR. A proposed novel function of the psoas minor revealed through cadaver dissection. Clin Anat. 2014;28(2):243–52.

Nunziata A, Blumenfeld I. Cadera a resorte: a proposito de una variedad. Prensa Med Argent. 1951;38(32):1997–2001.

Pelsser V, Cardinal E, Hobden R, et al. Extraarticular snapping hip: sonographic findings. AJR Am J Roentgenol. 2001;176(1):67–73.

Philippon MJ, Devitt BM, Campbell KJ, et al. Anatomic variance of the iliopsoas tendon. Am J Sports Med. 2014a;42(4):807–11.

Philippon MJ, Devitt BM, Campbell KJ, et al. Anatomic variance of the iliopsoas tendon. Am J Sports Med. 2014b;42(4):807–11.

Polster JM, Elgabaly M, Lee H, Klika A, Drake R, Barsoum W. MRI and gross anatomy of the iliopsoas tendon complex. Skelet Radiol. 2007;37(1):55–8.

Rajendran K. The insertion of the iliopsoas as a design favouring lateral rather than medial rotation at the hip joint. Singap Med J. 1989;30(5):451–2.

Rankin AT, Bleakley CM, Cullen M. Hip joint pathology as a leading cause of groin pain in the sporting population: a 6-year review of 894 cases. Am J Sports Med. 2015;43(7):1698–703.

Raya MA, Gailey RS, Gaunaurd IA, et al. Comparison of three agility tests with male servicemembers: Edgren side step test, t-test, and Illinois agility test. J Rehabil Res Dev. 2013;50:951–60.

Reurink G, Goudswaard GJ, Tol JL, et al. MRI observations at return to play of clinically recovered hamstring injuries. Br J Sports Med. 2014;48(18):1370–6.

Schaberg JE, Harper MC, Allen WC. The snapping hip syndrome. Am J Sports Med. 1984;12(5):361–5.

Serner A, Tol JL, Jomaah N, Weir A, Whiteley R, Thorborg K, Robinson M, Hölmich P. Diagnosis of acute groin injuries: a prospective study of 110 athletes. Am J Sports Med. 2015 Aug;43(8):1857–64.

Tatu L, Parratte B, Vuillier F, et al. Descriptive anatomy of the femoral portion of the iliopsoas muscle. Anatomical basis of anterior snapping of the hip. Surg Radiol Anat. 2001;23(6):371–4.

Winston P, Awan R, Cassidy JD, et al. Clinical examination and ultrasound of self reported snapping hip syndrome in elite ballet dancers. Am J Sports Med. 2006;35(1):118–26.

Witvrouw E, Danneels L, Asselman P, et al. Muscle flexibility as a risk factor for developing muscle injuries in male professional soccer players. Am J Sports Med. 2003;31:41–6.

Wunderbaldinger P, Bremer C, Matuszewski L, et al. Efficient radiological assessment of the internal snapping hip syndrome. Eur Radiol. 2001;11(9):1743–7.

Zambaldi M, Beasley I, Rushton A. Return to play criteria after hamstring muscle injury in professional football: a Delphi consensus study. Br J Sports Med. 2017;51:1221–6.

14 The Clinical Tests for RTT Decision-Making Process

14.1 Introduction

The identification of appropriate clinical tests depends on the type of muscle injury. Indeed, the specific biomechanical behavior of the different muscles requires the choice of specific tests. This chapter describes the clinical tests for RTT to be adopted in the case of:

1. Quadriceps injury.
2. Hamstring injury.
3. Adductor injury.
4. Soleus-gastrocnemius injury.
5. Short external hip rotator injury.
6. Iliopsoas injury.

Furthermore, the reader can access the videos of the various clinical tests using the dedicated QR code. Please note that RTT clinical tests are intended for a clinical judgment of successful healing. Therefore, all the tests described below are of medical competence. A positive result in the clinical tests of RTT allows the athlete to move on to the laboratory testing phase.

14.2 Quadriceps RTT Clinical Tests

The specific clinical test for RTT following quadriceps injury is the passive quadriceps stretch test (Witvrouw et al. 2003; Bouvier et al. 2017). The operator begins by testing the healthy limb and then moves on to the contralateral injured limb. The subject is lying on his side, and the operator stabilizes the hip with one hand. The operator first brings the patient's hip to maximum extension and then flexes his knee as much as possible, bringing the leg to the thigh (Fig. 14.1). Once the position is reached, a second operator measures both hip extension and knee flexion (Fig. 14.2).

The reference criteria (passing requirements) are:

1. No pain on the VAS scale.
2. Elongation equal to the contralateral limb (maximum tolerable difference equal to 10%).

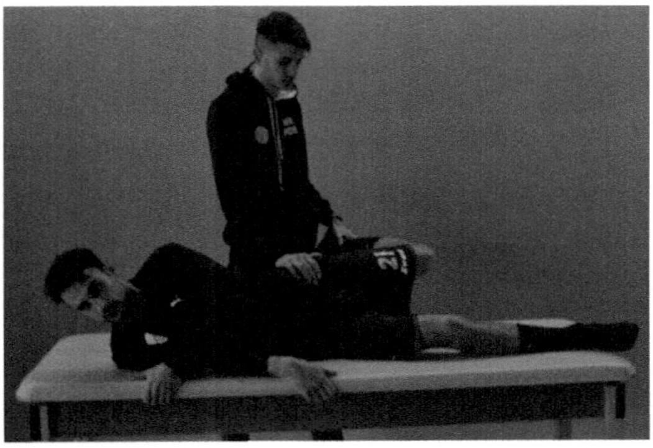

Fig. 14.1 Passive quadriceps stretch test final position

G. N. Bisciotti et al., *Return-to-Play after Lower Limb Muscle Injury in Football*,
https://doi.org/10.1007/978-3-030-84950-4_14

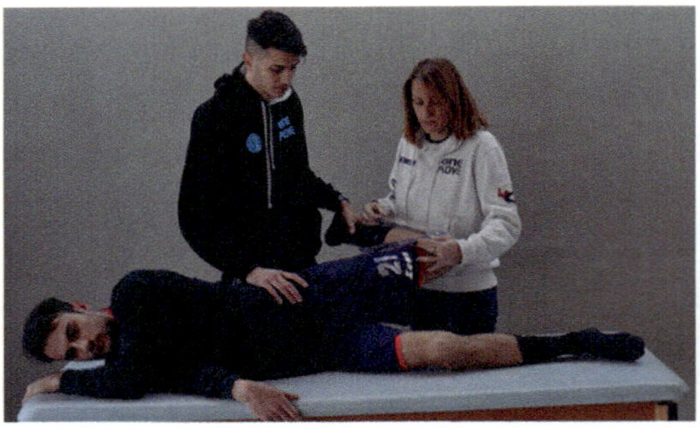

Fig. 14.2 Passive quadriceps stretch test measure

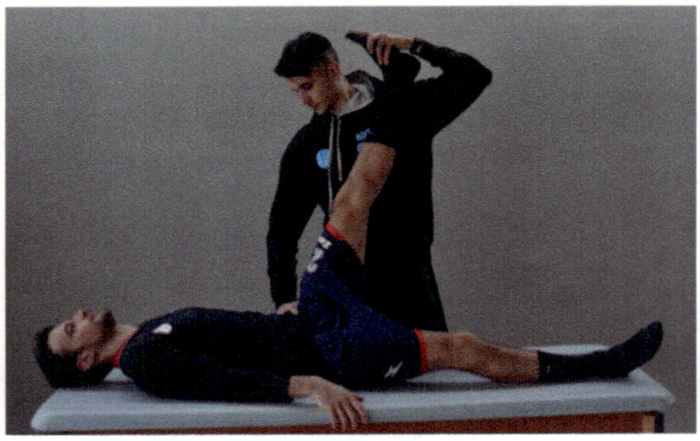

Fig. 14.3 Passive straight leg raise test final position

14.3 Hamstring RTT Clinical Tests

The specific clinical tests for RTT following hamstring injury are the passive straight leg raise test (Delvaux et al. 2014; Kellis et al. 2015; Ridehalgh et al. 2015) and the dynamic flexibility H test (Askling et al. 2010).

14.3.1 Straight Leg Raise Test

During the passive straight leg raise test, the patient is on a bench in a supine position. The test starts with the healthy limb and then moves on to the contralateral injured limb. The operator with one hand blocks the contralateral hip while bringing with the other hand the other leg to maximum hip flexion keeping the knee extended and the foot in a neutral position (Fig. 14.3). Once the position is reached, a second operator measures both hip extension and knee flexion (Fig. 14.4).

The reference criteria are:

1. No pain on the VAS scale.
2. Elongation equal to the contralateral limb (maximum tolerable difference equal to 10%).

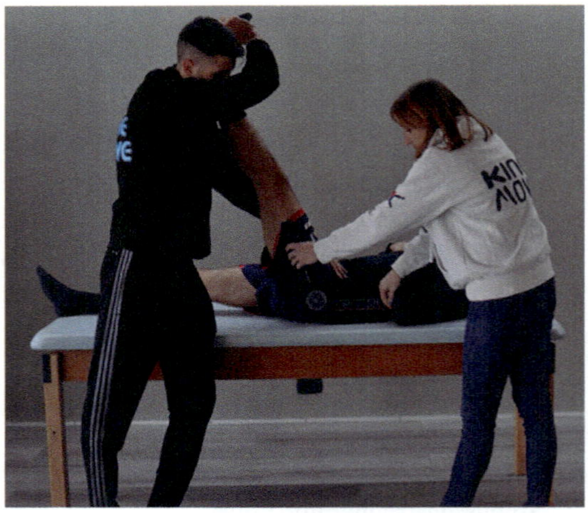

Fig. 14.4 Passive straight leg raise test measure

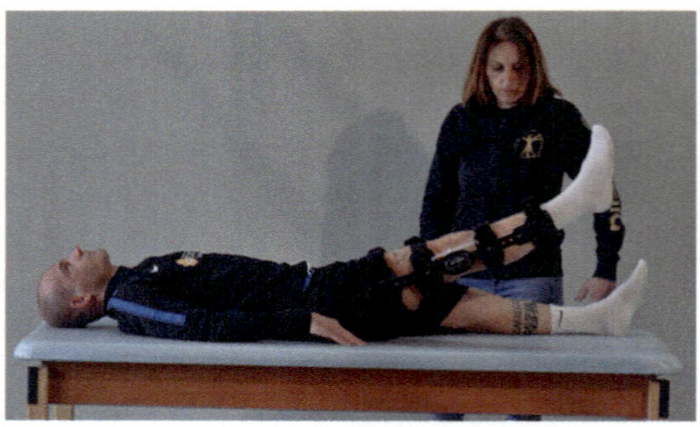

Fig. 14.5 For the dynamic flexibility H test. The subject is positioned on a bench in a supine position wearing the brace and with the axis of the electrogoniometer aligned with the axis of the hip joint

14.3.2 Dynamic Flexibility H Test

To perform the dynamic flexibility H test (Askling et al. 2010), a brace blocking the knee joint is necessary while an electrogoniometer is optional. The subject is positioned on a bench in a supine position wearing the brace and with the axis of the electrogoniometer aligned with the axis of the hip joint (Fig. 14.5). The knee brace ensures full knee extension of the tested leg. The healthy limb is tested first, followed by the injured contralateral limb. During the test, the foot of the tested leg is kept in a slightly

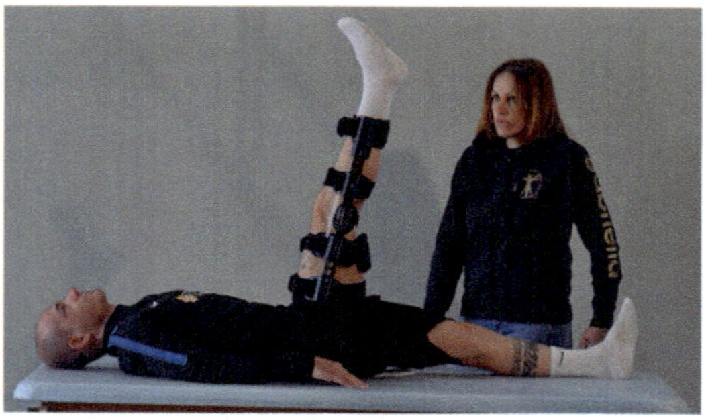

Fig. 14.6 During the test the subject performs a straight leg raise as fast as possible to the highest point without taking any risk of injury

plantar-flexed position; the subject instructed to perform a straight leg raise as fast as possible to the highest point without taking any risk of injury (Fig. 14.6). A set of three consecutive trials is performed, preceded by one practice trial with submaximal effort. After the three active trials, the subject estimates the experience of insecurity on a "feel of insecurity scale" from 0 to 100 and the pain on a VAS scale from 0 to 10. Flexibility is measured as the largest range of motion (ROM) of three consecutive trials, calculated from the electrogoniometer data, while mean angular velocity is calculated from the electrogoniometer data recorded during the test. The trial with the largest ROM is used for analysis. If the test is performed with the use of an electrogoniometer, the reference criteria are:

1. The ROM difference between the two limbs is <10%.
2. The angular velocity difference between the two limbs is <10%.
3. The feeling of insecurity is absent on a "feel of insecurity scale" from 0 to 100.
4. No pain on the VAS scale.

If the test is performed without the use of an electrogoniometer, only the feeling of insecurity and the pain on VAS scale are considered.

14.4 Adductors RTT Clinical Tests

The specific clinical tests for RTT following adductor muscle injury are the pubic stress test (Hogan and Lovell 1998), the resisted hip adduction test (Hogan and Lovell 1998; Engebretsen et al. 2010; Bisciotti 2013), the squeeze test (Delahunt et al. 2011a; Delahunt et al. 2011b; Nevin and Delahunt 2014; Hodgson et al. 2015), and the adductor passive test (Atkinson et al. 2010; Witvrouw et al. 2003).

14.4.1 Pubic Stress Test

During the pubic stress test (Hogan and Lovell 1998), the subject is positioned on a bench in a supine position and with the two legs outside the bed edge. The healthy limb is tested first, followed by the injured contralateral limb. The operator performs a hip passive flexion of the healthy limb and an extension-abduction of the injured limb (Fig. 14.7). Once the position is reached, a second operator measures the injured limb degrees of the abduction (Fig. 14.8).

The reference criteria are:

1. No pain on the VAS scale.
2. Difference of elongability in abduction of the injured limb <10% compared to the healthy limb.

14.4.2 Resisted Hip Adduction Test

During the resisted hip adduction test (Hogan and Lovell 1998; Engebretsen et al. 2010; Bisciotti

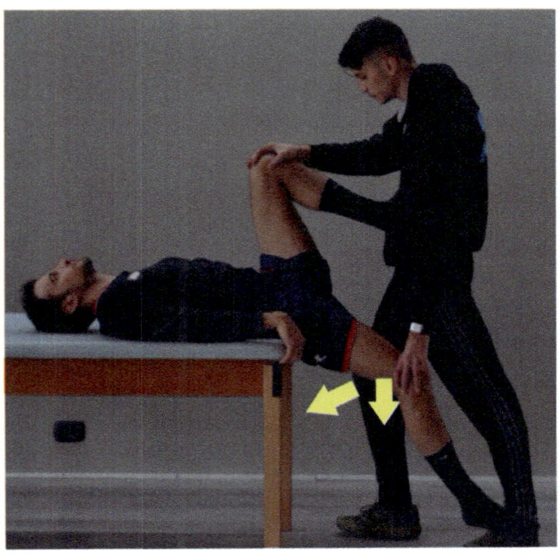

Fig. 14.7 Pubic stress test final position

Fig. 14.8 Injured limb abduction degree measurement

2013), the subject is positioned on a bench in a supine position and with the two legs outside the bed edge. The healthy limb is tested first, followed by the injured contralateral limb. A handle dynamometer is required to perform the test. The operator performs a complete flexion of the healthy limb while asking the patient for a maximum isometric adduction of the injured limb (Fig. 14.9). The reference criteria are:

1. No pain on the VAS scale.
2. The strength difference between non-injured and injured legs must be <10%.

14.4.3 The Squeeze Test

The squeeze test (Delahunt et al. 2011a; Delahunt et al. 2011b; Nevin and Delahunt 2014; Hodgson et al. 2015) is performed in four different positions:

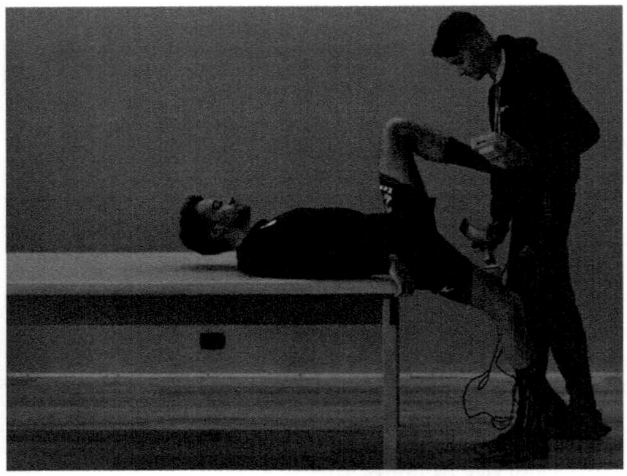

Fig. 14.9 During the resisted hip adduction test, the healthy limb is in full flexion, and the patient performs a maximum isometric adduction of the injured limb

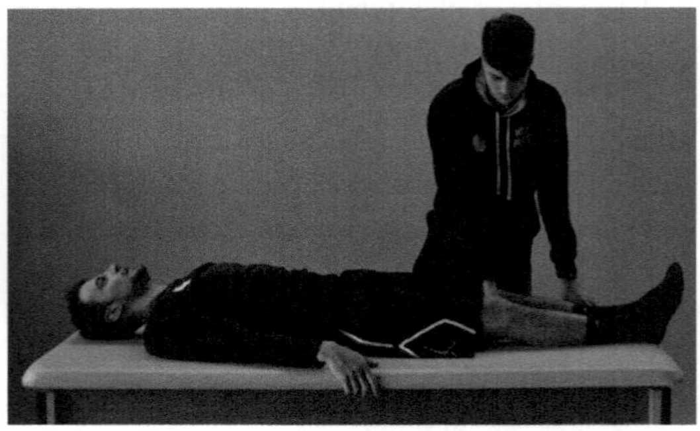

Fig. 14.10 Squeeze test with proximal resistance

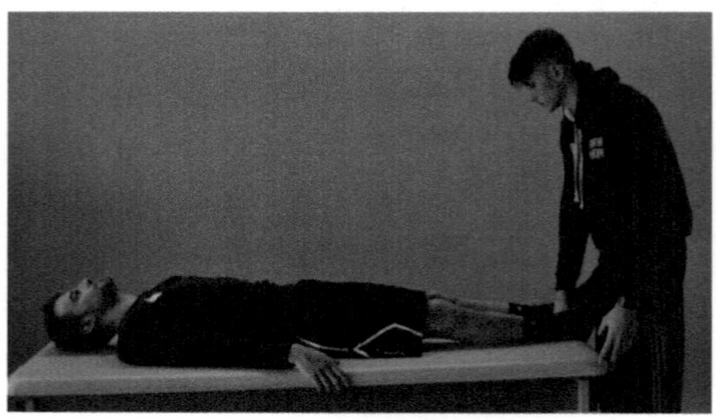

Fig. 14.11 Squeeze test with distal resistance

1. With proximal resistance (Fig. 14.10).
2. With distal resistance (Fig. 14.11).
3. With distal resistance and abducted legs (Fig. 14.12).
4. With proximal resistance and flexed legs (Fig. 14.13).

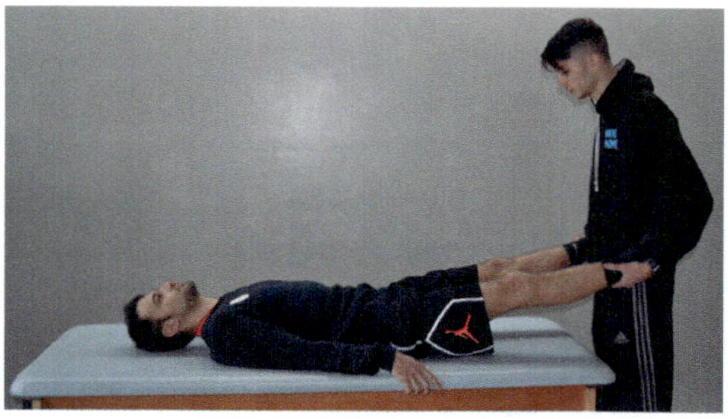

Fig. 14.12 Squeeze test with distal resistance and abducted legs

Fig. 14.13 Squeeze test with proximal resistance and flexed legs (specific for gracilis muscle)

During the test, the patient is on a bench in a supine position with extended legs, except during the specific test for the gracilis muscle, during which the subject has his legs flexed (Fig. 14.13). The operator asks to the subject to perform a maximal isometric contraction for about 5 s. The reference criteria is no pain on the VAS scale.

Furthermore, it is possible to perform a variant of the test with the use of a dynamometer (Bisciotti et al. 2016a, b; Bisciotti et al. 2019) as shown in Fig. 14.14. The healthy leg is tested before the injured leg. The reference criteria are:

1. No pain on the VAS scale.
2. The strength difference between non-injured and injured legs is <10%.

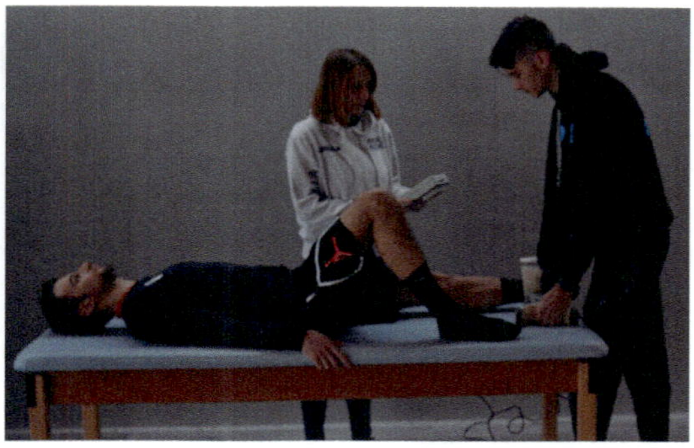

Fig. 14.14 Monopodalic squeeze test with the use of a dynamometer

14.5 Soleus-Gastrocnemius RTT Clinical Tests

The specific clinical tests for RTT following soleus-gastrocnemius injury are the weight bearing lunge test (Hoch and McKeon 2011; Powden et al. 2015; Baumach et al. 2016) and the heel-raise test (Moller et al. 2005; Harris-Love et al. 2014).

14.5.1 Weight Bearing Lunge Test

The test start with the patient kneeling in front of a wall. The healthy leg is tested first, followed by the injured leg. The first test is performed with a distance between the tip of the big toe and the wall equal to 5 cm. From this position, the patient must to be able to touch the wall with the knee without lifting the heel off the ground (Fig. 14.15). The distance is progressively increased, and the test is performed several times. The best result is recorded. The reference criteria are:

1. No pain on the VAS scale.
2. Difference in elongability of the injured limb <10% compared to the healthy limb.

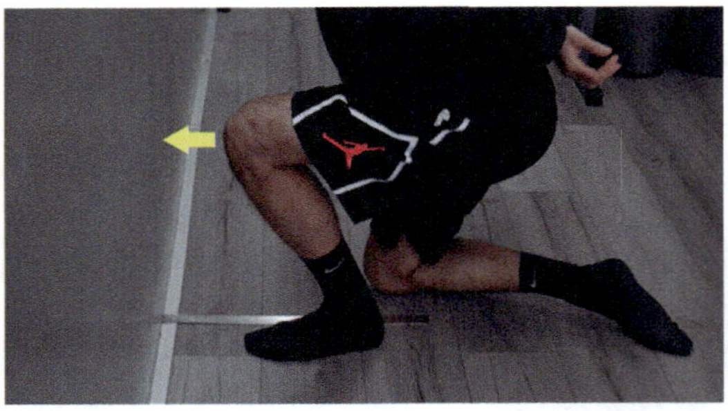

Fig. 14.15 The weight bearing lunge test

14.5.2 Heel-Raise Test

Starting position: the subject is barefoot, using their fingers supported on a wall for balance, with the elbows slightly flexed, keeping the spine in a neutral position. The feet are slightly apart (hip width).

Test execution: the subject raises their heel as high as possible with their knee full extended while the rater defines the range of motion using a square tool (Fig. 14.16). After the amplitude definition, the test is first performed with the non-injured limb. At the signal of the rater, the subject raises their heel vertically up to the maximum height possible and lowers it completely to the ground, performing the highest possible number of repetitions in a predefined period of 30 s. The subject should touch the square with their head whenever the predefined height is reached and touch the ground with their heel at the end of each movement. The score is represented by the number of correct movements performed in 30 s.

The reference criteria are:

1. No pain on the VAS scale.
2. The difference between the number of correct movements performed in 30 s with the injured limb and those performed with the healthy limb must not exceed 10%.

Fig. 14.16 Heel-raise test

14.6 Short External Hip Rotator Injury RTT Clinical Tests

The specific assessment for short external hip rotator muscle injury is based on the following tests (Delp et al. 1999; Busfield and Romero 2009; Valente et al. 2011; Delin et al. 2017):

1. Pace and Nagle maneuvers test.
2. Beatty maneuver test.
3. Freiberg maneuver test.
4. Internal rotation test.

14.6.1 Pace and Nagle Maneuver Tests

During the Pace and Nagle maneuver tests (Fig. 14.17), the patient is in seated position on the edge of the bed. The operator asks the patient to perform an isometric maximal hip abduction under resistance. The test may be performed also in monopodalic execution with the use of a dynamometer (Fig. 14.18). In the case of monopodalic execution, the non-injured leg is tested first. The reference criteria are:

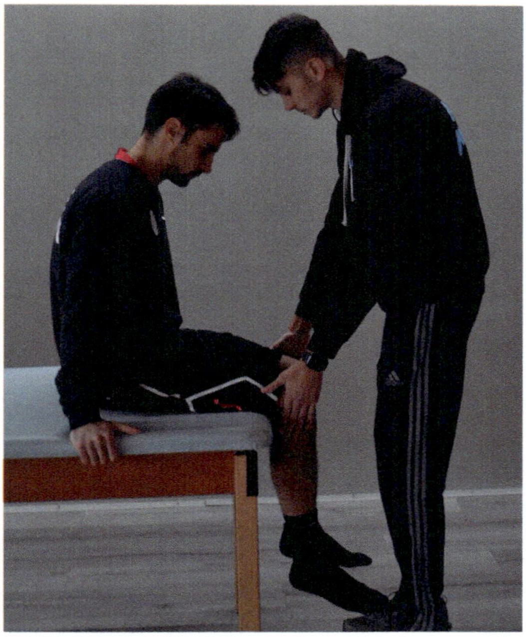

Fig. 14.17 Pace and Nagle maneuver tests

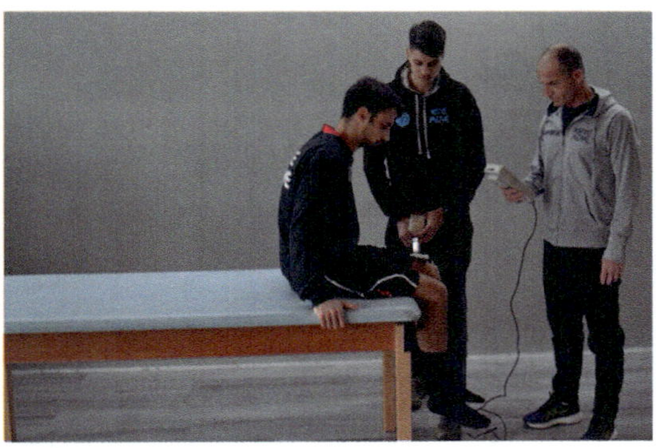

Fig. 14.18 Pace and Nagle maneuver tests in monopodalic execution with dynamometric evaluation

1. No pain on the VAS scale.
2. The strength difference between non-injured and injured legs must be <10% (in case of dynamometric evaluation).

14.6.2 Beatty Maneuver Test

In the Beatty maneuver test (Fig. 14.19), the patient is in lateral lying position, injured side up, superior foot behind inferior popliteal fossa of the uninjured limb, and hip in a position of flexion-adduction-internal rotation. The non-injured leg is tested first. The operator asks the patient to raise the knee against resistance. The test may be performed also in monopodalic execution with the use of a dynamometer (Fig. 14.20). The reference criteria are:

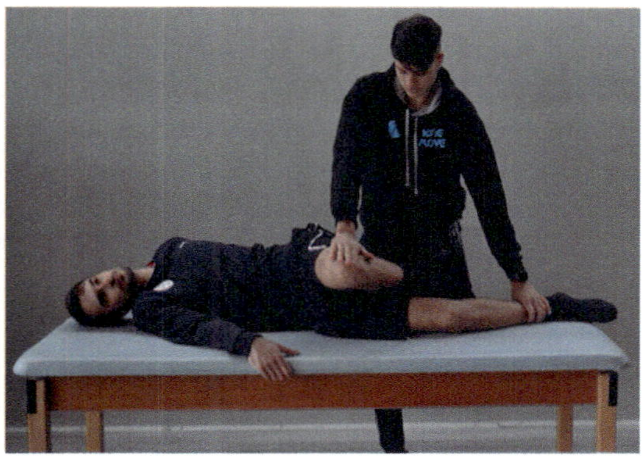

Fig. 14.19 The Beatty maneuver test

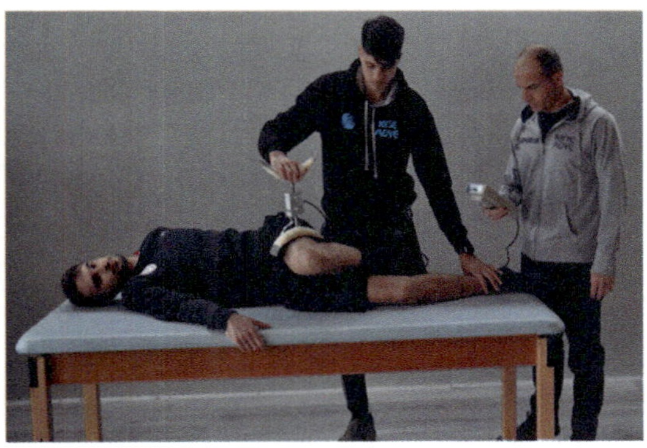

Fig. 14.20 The Beatty maneuver test with dynamometric evaluation

1. No pain on the VAS scale.
2. The strength difference between non-injured and injured legs must be <10% (in case of dynamometric evaluation).

14.6.3 Freiberg Maneuver Test

During the Freiberg maneuver test (Fig. 14.21), the patient is in dorsal decubitus position. The non-injured leg is tested first. The operators bring the subject hip to a position of flexion-adduction-medial rotation. The reference criteria is no pain on VAS scale.

14.6.4 Internal Rotation Test

In the internal rotation test, the patient is sitting on a table with his feet beyond the edge. The non-injured leg is tested first. The operator, keeping the patient's leg in extension, performs an internal rotation of the hip (Fig. 14.22). Once the maximum internal rotation position has been reached, a second operator measures the internal rotation degrees (Fig. 14.23). The reference criteria are:

1. No pain on the VAS scale.
2. Intra-rotation equal to the contralateral limb (maximum tolerable difference equal to 10%).

Fig. 14.21 The Freiberg maneuver test

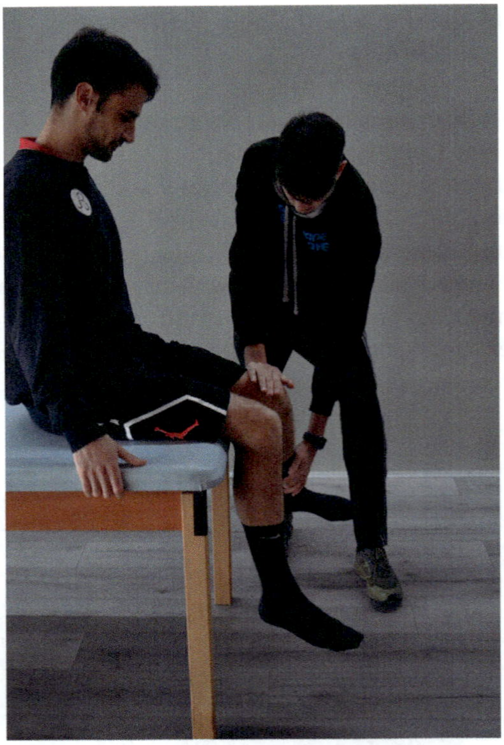

Fig. 14.22 Internal rotation test

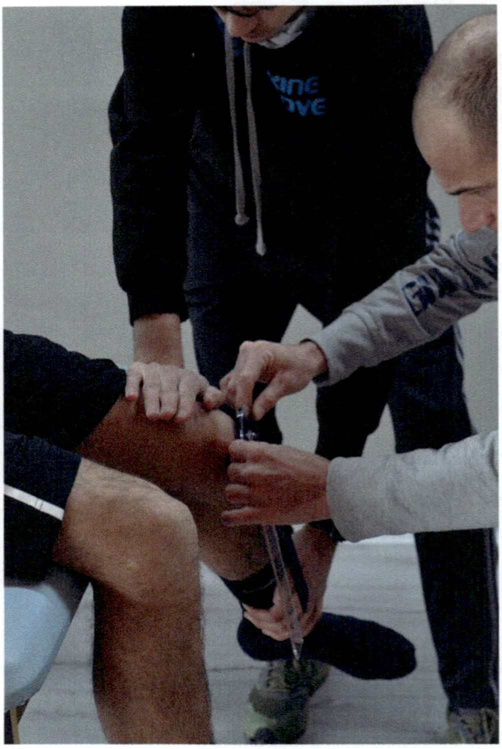

Fig. 14.23 Internal rotation test ROM measurement

14.7 Iliopsoas Injury RTT Clinical Tests

The specific assessment for iliopsoas injury is the modified Thomas test (Harvey 1998; Ferber et al. 2010) performed in the following modalities:

1. In passive stretching.
2. In isometric contraction (the use of a dynamometer is recommended).
3. In concentric contraction (the use of a dynamometer is recommended).
4. In eccentric contraction (the use of a dynamometer is recommended).

During the modified Thomas test executed in passive stretching, the patient is supine on the table, avoiding a posterior tilt of the pelvis. About 50% of the pelvis must protrude from the edge of the table (i.e., the patient is balanced at the edge of the table). The first operator maximally flexes the contralateral leg of the patient and performs a maximal stretching of the opposite injured leg (Fig. 14.24a), while a second operator measures the ROM (Fig. 14.24b).

The reference criteria are:

1. No pain on the VAS scale
2. Elongation equal to the contralateral limb (maximum tolerable difference equal to 10%).

During the modified Thomas test executed in isometric contraction (Fig. 14.25), the patient is in the same position described for the previous test. The first operator maximally flexes the contralateral leg of the patient, while a second operator asks the patient to perform with the injured leg a maximal isometric contraction. The use of a handle dynamometer to measure the produced strength is recommended.

The reference criteria are:

1. No pain on the VAS scale.
2. The strength difference between non-injured and injured legs must be <10% (in case of dynamometric evaluation).

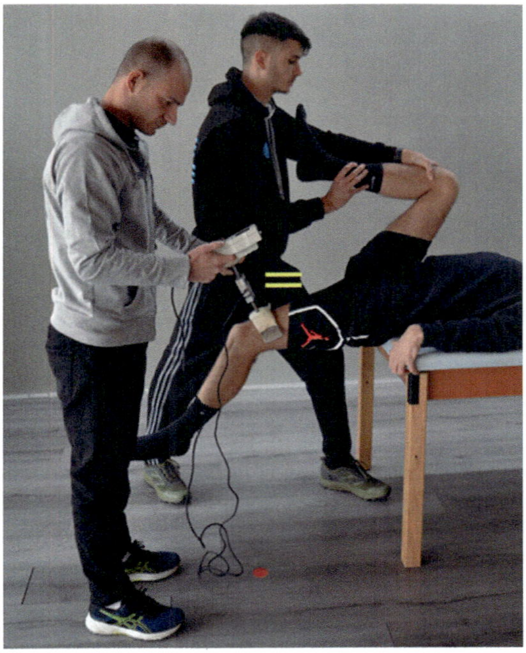

Fig. 14.25 Modified Thomas test executed in isometric contraction

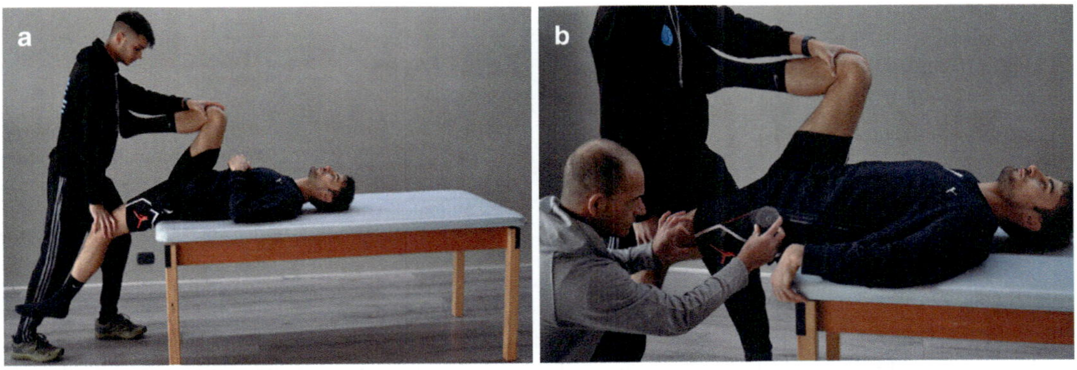

Fig. 14.24 Modified Thomas test is executed in passive stretching (**a**), while a second operator measures the ROM (**b**)

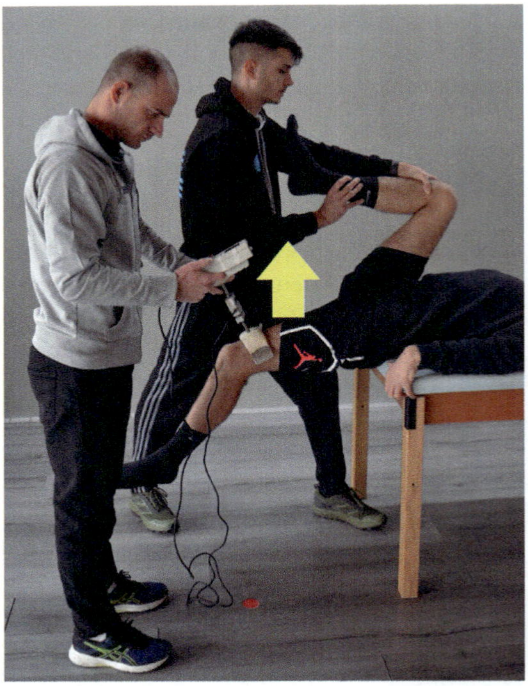

Fig. 14.26 Modified Thomas test executed in concentric contraction. Yellow arrow: force exerted by the patient

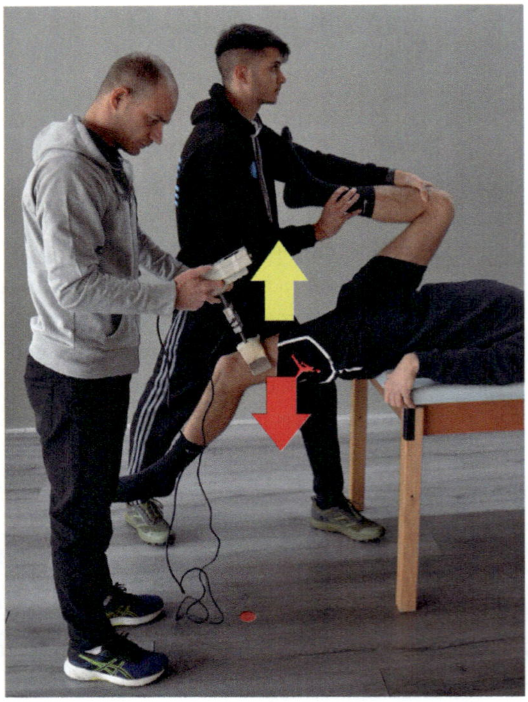

Fig. 14.27 Modified Thomas test executed in eccentric contraction. Yellow arrow: force exerted by the patient; red arrow: force exerted by the operator

During the modified Thomas test executed in concentric contraction (Fig. 14.26), the patient is in the same position described for the previous tests. The first operator maximally flexes the contralateral leg of the patient, while a second operator asks the patient to perform with the injured leg a series of four to six maximal concentric contractions. The use of a handle dynamometer to measure the produced strength is recommended.

The reference criteria are

1. No pain on the VAS scale.
2. The strength difference between non-injured and injured legs must be <10% (in case of dynamometric evaluation).

During the modified Thomas test executed in eccentric contraction (Fig. 14.27), the patient is in the same position described for the previous tests. The first operator maximally flexes the contralateral leg of the patient, while a second operator asks the patient to perform with the injured leg a series of four to six maximal eccentric contractions. The use of a handle dynamometer to measure the produced strength is recommended.

The reference criteria are:

1. No pain on the VAS scale.
2. The strength difference between non-injured and injured legs must be <10% (in case of dynamometric evaluation).

References

Askling CM, Nilsson J, Thorstensson A. A new hamstring test to complement the common clinical examination before return to sport after injury. Knee Surg Sports Traumatol Arthrosc. 2010;18:1798–803.

Atkinson HDE, Johal P, Falworth MS, et al. Adductor tenotomy: its role in the management of sports-related chronic groin pain. Arch Orthop Trauma Surg. 2010;130:965–70.

Baumach SF, Braunstein M, Regauer M, et al. Diagnosis of Musculus gastrocnemius tightness - key factors for the clinical examination. J Vis Exp. 2016;113

Bisciotti GN. La tendinopatia degli adduttoria nel calciatore quando il ritorno alla corsa? Strength Condition. 2013;5:11–6.

Bisciotti GN, Quaglia A, Belli A, et al. Return to sports after ACL reconstruction: a new functional test protocol. Muscles Ligaments Tendons J. 2016a;06:499–509.

Bisciotti GN, Volpi P, Zini R, et al. Groin pain syndrome Italian consensus conference on terminology, clinical evaluation and imaging assessment in groin pain in athlete. BMJ Open Sport Exerc Med. 2016b;2:e000142.

Bisciotti GN, Volpi P, Alberti G, et al. Italian consensus statement (2020) on return to play after lower limb muscle injury in football (soccer). BMJ Open Sport Exerc Med. 2019;5:e000505.

Bouvier T, Opplert J, Cometti C, et al. Acute effects of static stretching on muscle-tendon mechanics of quadriceps and plantar flexor muscles. Eur J Appl Physiol. 2017;117:1309–15.

Busfield BT, Romero DM. Obturator internus strain in the hip of an adolescent athlete. Am J Orthop (Belle Mead NJ). 2009 Nov;38(11):588–9.

Delahunt E, Kennelly C, McEntee BL, et al. The thigh adductor squeeze test: 45° of hip flexion as the optimal test position for eliciting adductor muscle activity and maximum pressure values. Man Ther. 2011b;16:476–80.

Delahunt E, McEntee BL, Kennelly C, et al. Intra-rater reliability of the adductor squeeze test in gaelic games athletes. J Athl Train. 2011a;46:241–5.

Delin C, Vandensteen JY, Roger B. Hip Short External Rotator Muscles Injuries. In: Roger B, Guermazi A, Skaf A, editors. Muscle injuries in sport athletes. Cham: Springer Editions; 2017.

Delp SL, Hess WE, Hungerford DS, Jones LC. Variation of rotation moment arms with hip flexion. J Biomech. 1999;32(5):493–501.

Delvaux F, Rochcongar P, Bruyere O, et al. Return-to-play criteria after hamstring injury: actual medicine practice in professional soccer teams. J Sports Sci Med. 2014;13:721–3.

Engebretsen AH, Myklebust G, Holme I, et al. Intrinsic risk factors for groin injuries among male soccer players: a prospective cohort study. Am J Sports Med. 2010;38:2051–7.

Ferber R, Kendall KD, McElroy LJ. Normative and critical criteria for iliotibial band and iliopsoas muscle flexibility. Athl Train. 2010 Jul-Aug;45(4):344–8.

Harris-Love MO, Shrader JA, Davenport TE, et al. Are repeated single-limb heel raises and manual muscle testing associated with peak plantar-flexor force in people with inclusion body myositis? Phys Ther. 2014;94:543–52.

Harvey D. Assessment of the flexibility of elite athletes using the modified Thomas test. Br J Sports Med. 1998 Mar;32(1):68–70.

Hoch MC, McKeon PO. Normative range of weight-bearing lunge test performance asymmetry in healthy adults. Man Ther. 2011 Oct;16(5):516–9.

Hodgson L, Hignett T, Edwards K. Normative adductor squeeze tests scores in rugby. Phys Ther Sport. 2015;16:93–7.

Hogan A, Lovell G. The groin pain provocation test. In: Brown A, editor. 4thWorld football symposium conference proceedings. London: Routledge; 1998.

Kellis E, Ellinoudis A, Kofotolis N. Hamstring elongation quantified using ultrasonography during the straight leg raise test in individuals with low back pain. PM R. 2015;7:576–83.

Moller M, Lind K, Styf J, et al. The reliability of isokinetic testing of the ankle joint and a heel-raise test for endurance. Knee Surg Sports Traumatol Arthrosc. 2005;13:60–71.

Nevin F, Delahunt E. Adductor squeeze test values and hip joint range of motion in Gaelic football athletes with longstanding groin pain. J Sci Med Sport. 2014;17:155–9.

Powden CJ, Hoch JM, Hoch MC. Reliability and minimal detectable change of the weight-bearing lunge test: A systematic review. Man Ther. 2015 Aug;20(4):524–32.

Ridehalgh C, Moore A, Hough A. Sciatic nerve excursion during a modified passive straight leg raise test in asymptomatic participants and participants with spinally referred leg pain. Man Ther. 2015;20:564–9.

Valente HC, Marques FO, Da Silva De Souza L, Abib RT, Ribeiro DC. Injury of the external obturator muscle in professional soccer athletes. Rev Bras Med Esporte. 2011;17(1):36–9.

Witvrouw E, Danneels L, Asselman P, et al. Muscle flexibility as a risk factor for developing muscle injuries in male professional soccer players. Am J Sports Med. 2003;31:41–6.

15 The Laboratory Tests for RTT Decision-Making Process

15.1 Introduction

The laboratory tests for the RTT decision-making process consist of a series of dynamometric tests that can be performed with the aid of handle dynamometers, computerized specific devices, isokinetic devices, and force platforms. This chapter describes the laboratory tests suggested for each different muscle injury:

1. Quadriceps injury.
2. Hamstring injury.
3. Adductor injury.
4. Soleus-gastrocnemius injury.
5. Short external hip rotator injury.

Concerning the basic principles of dynamometric tests, see Chap. 6 for further details.

15.2 Quadriceps RTT Laboratory Tests

The specific laboratory tests for RTT following quadriceps injury are:

1. Isometric tests (Bisciotti 1998; Delvaux et al. 2014).
2. Isotonic tests (Bisciotti 2003).
3. Isokinetic test (Menzel et al. 2013; Sanfilippo et al. 2013).
4. Synhcro plates test (Bisciotti et al. 2016; Bisciotti et al. 2019).

15.2.1 Isometric Tests

Isometric tests may be performed either to a normal leg extension machine, equipped with a load cell (Fig. 15.1) or with a specific device (Kineo Globus, Italy) in isometric modality (Fig. 15.2). We suggest to perform the test with two different knee standard flexion angles of 45° and 70°, respectively, since quadriceps voluntary activation deficits are significantly higher with 45° knee flexion in comparison to greater or lesser angles (Chandramouli and Theuerkauf 2015), while the quadriceps maximal strength is reached at 70° knee flexion (Knapik et al. 1983). The test must be done with the non-injured leg first. The parameters to be taken into consideration are the maximum force value and the rate of force development value (Wilson et al. 1995; Verchoshansky 1996). This test is particularly effective in the case of middle third injuries or for distal injuries, but it is not reliable for the proximal injuries. For the latter, the isometric test must be performed with the leg in extended position, with a load cell (Fig. 15.3), or a handle dynamometer (Fig. 15.4), or a specific device (Kineo Globus, Italy) in isometric modality (Fig. 15.5).

G. N. Bisciotti et al., *Return-to-Play after Lower Limb Muscle Injury in Football*,
https://doi.org/10.1007/978-3-030-84950-4_15

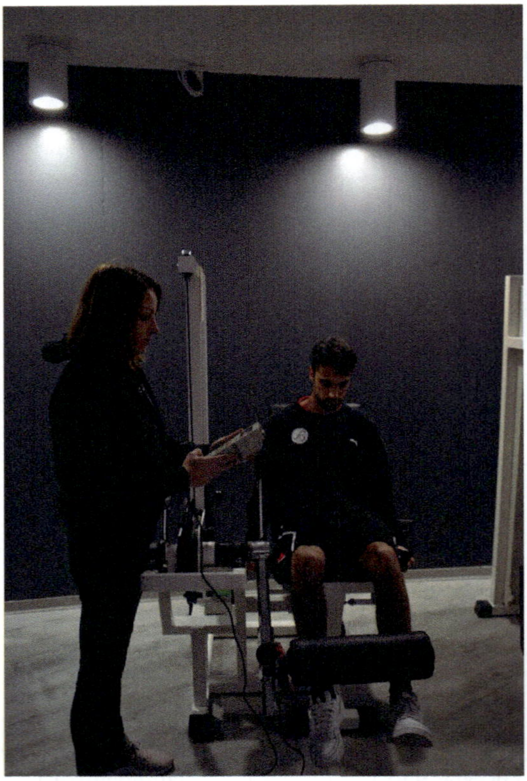

Fig. 15.1 Isometric test performed with the use of a leg extension equipped with a load cell

Fig. 15.3 Isometric test performed with the use of a load cell in the case of quadriceps proximal injury

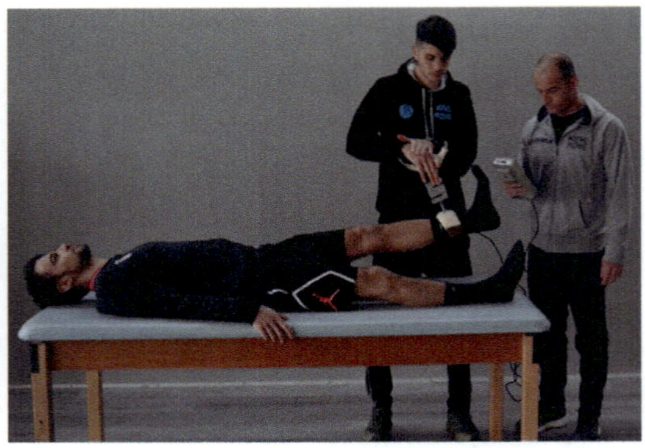

Fig. 15.4 Isometric test with the use of a handle dynamometer in the case of quadriceps proximal injury

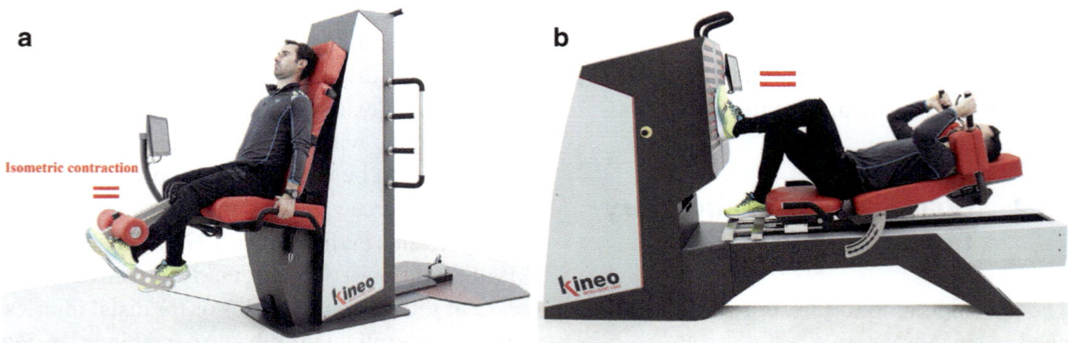

Fig. 15.2 Isometric test performed with the use of a specific device (Kineo globus, Italy) in open kinetic chain (**a**) and in closed kinetic chain (**b**)

The test must be done with the non-injured leg first. The reference values (passing requirements) are:

1. No pain on VAS scale during the test.

The dynamometric values of the injured leg differ from pre-injury values or from healthy limb values by a maximum of 10%.

15.2.2 Isotonic Tests

The isotonic test may be performed with the use of a specific device (Kineo Globus, Italy) in isotonic modality (Fig. 15.6). The test must be done with the non-injured leg first. Here too, the test is effective in the case of middle third injuries or for distal injuries but is not reliable for the proximal injuries. In the case of proximal injuries, the isotonic test must be performed with the same device (Kineo Globus, Italy) but in another testing position as shown in Fig. 15.7.

The parameters to be taken into consideration are the maximum force value and the peak of power production (Fig. 15.8) (Driss et al. 1998; Jaskólska et al. 1999).The reference values are:

1. No pain on VAS scale during the test.
2. All the calculated dynamometric values of the injured leg differ from the pre-injury values or from the healthy limb values by a maximum of 10%.

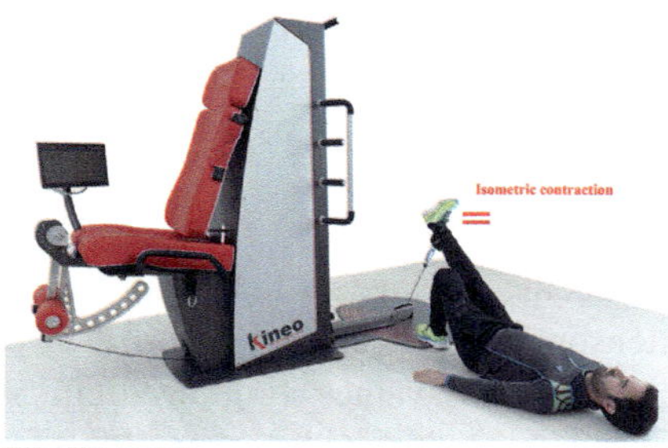

Fig. 15.5 Isometric test performed with the use of a specific device (Kineo Globus, Italy) in the case of proximal injury

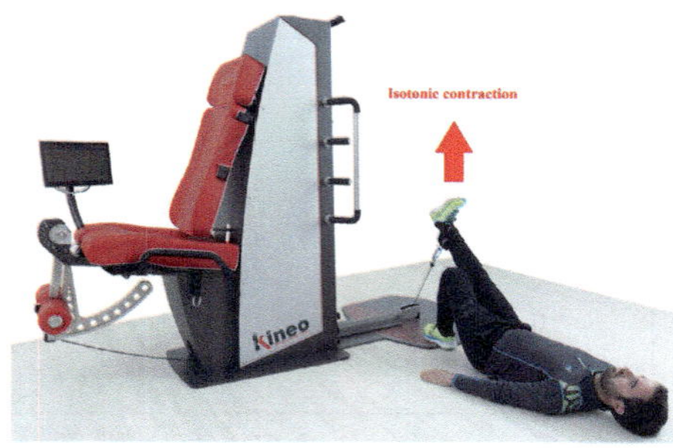

Fig. 15.7 Isotonic test performed with the use of a specific device (Kineo Globus, Italy) in isotonic modality in the case of quadriceps proximal injury

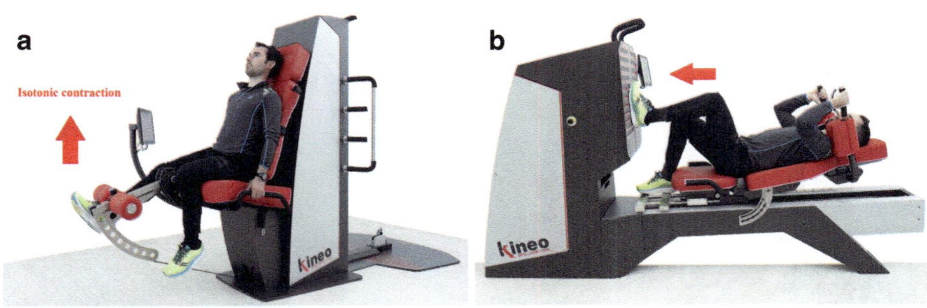

Fig. 15.6 Isotonic test performed with the use of a specific device (Kineo Globus, Italy) in isotonic modality in the case of quadriceps middle third or distal injury. The test is performed in open kinetic chain (**a**) and in closed kinetic chain (**b**)

A particular case of isotonic test is the synchro plates test (Bisciotti et al. 2016).

During synchro plates test, the subject is on a double synchronized force platform (Thesis System, Globus Italy. Reliability of the measure: 0.5 Kg; error of the device: <1%; sample rate 1000 Hz; sample depth 140 bit) and performs a countermovement jump (CMJ) with 90° flexion knee angle. During the landing phase, the subject stopped the movement back at 90° flexion knee angle (Fig. 15.9). The force platforms, after recording simultaneously the strength signal, automatically calculated and compared the two force signals recorded in four different phases, specifically:

- First phase or negative overstretching phase.
- Second phase or maximal negative strength phase.
- Third phase or maximal positive strength phase.
- Fourth phase or landing phase.

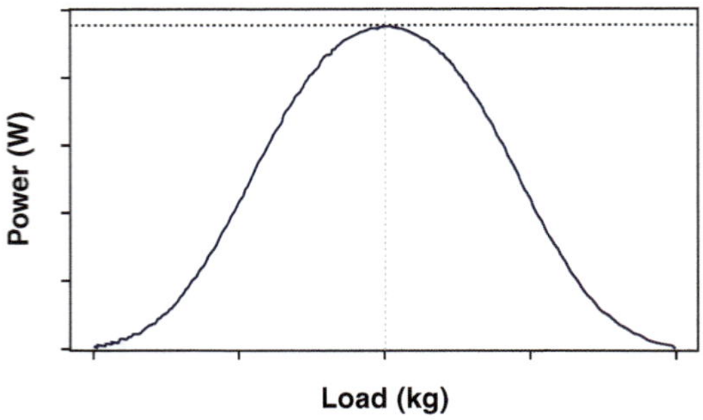

Fig. 15.8 During the isotonic test the power production in relationship with the load may be calculated

During the above-listed four phases, corresponding to different muscular activation patterns, the following parameters are calculated:

1. The maximal negative acceleration value calculated during the stretching phase (MAneg recorded during the first phase).
2. The maximal negative force value (MFecc recorded during the second phase).
3. The maximal positive force value (MFcon recorded during the third phase).
4. The maximal landing force (or impact force) value (MFimp recorded during the fourth phase).

The biomechanical interpretation of the above parameters recorded during the different phases can be briefly summarized as follows:

1. MAneg (Fig. 15.10) and MFecc (Fig. 15.11) are specific parameters regarding the muscle extensor apparatus negative phase. The difference between values (injured limb versus healthy limb) must be less than 10%.
2. MFcon (Fig. 15.12) refers to the phase when the elastic energy stored during the negative phase is transformed into mechanical work.

Fig. 15.9 The synchro plates test protocol provides that the subject performs a CMJ (90° knee flexion angle) and stops the landing phase with a knee angle back to 90°. Two synchronized force platforms (Thesis System, Globus Italy. Reliability of the measure: 0.5 Kg; error of device

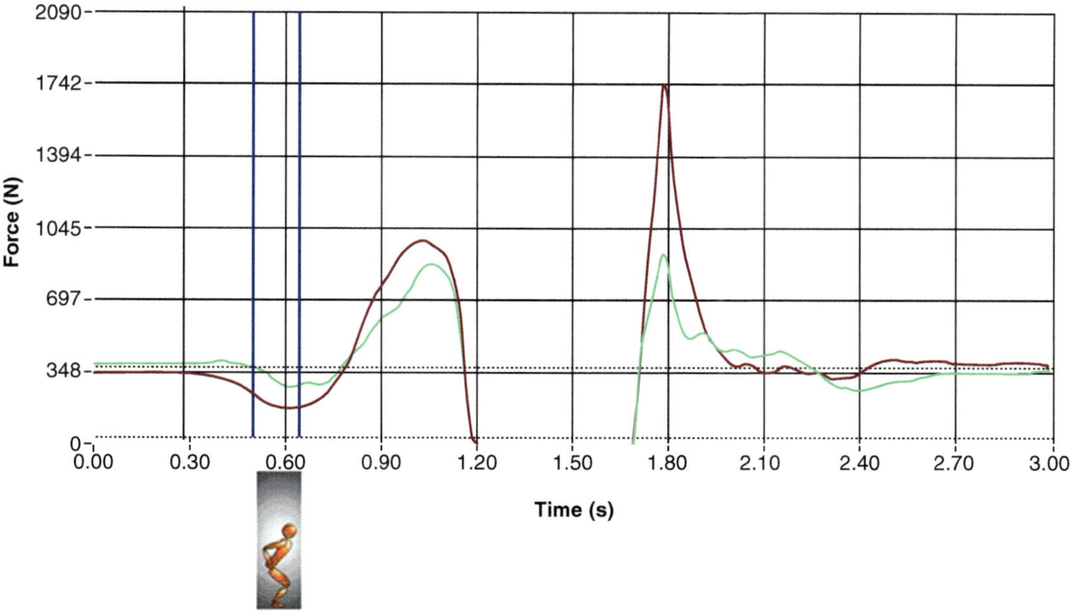

Fig. 15.10 The negative acceleration peak (MAneg) represents the overstretching capabilities of the extensor muscles. Red line: uninjured limb; green line: injured limb

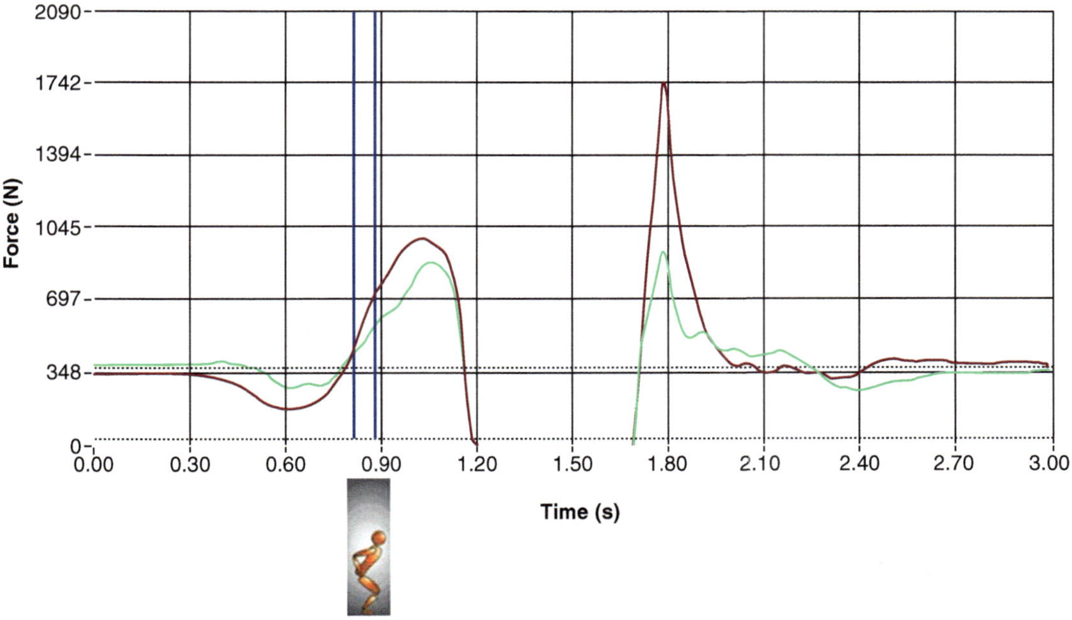

Fig. 15.11 The maximum force applied during the negative phase (MFecc) represents the maximum negative strength expressed by the extensor muscles immediately before the inversion of the movement (i.e., immediately before the positive phase). Red line: uninjured limb; green line: injured limb

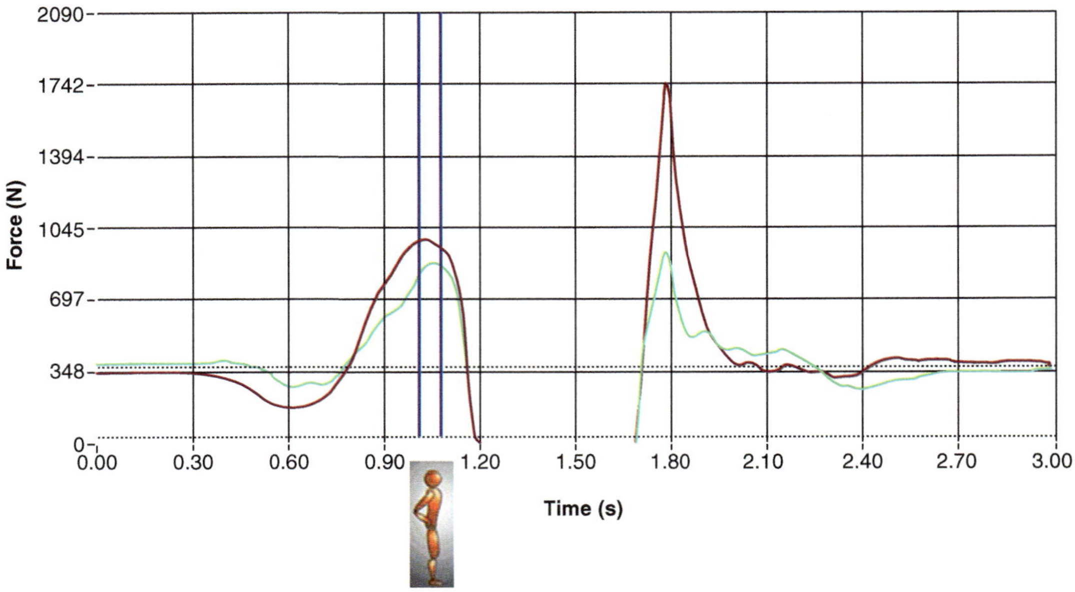

Fig. 15.12 The MFcon value represents the capacity of storage and reuse of elastic energy by the extensors muscle-tendon unit. Red line: uninjured limb; green line: injured limb

The mechanical work generated from the stored elastic energy enhances the force produced by the muscle contractile component during the positive phase. Therefore, the value recorded during this phase represents the extensor muscles elastic behavior. The difference between values (injured limb versus healthy limb) must be less than 10%.

3. MFimp (Fig. 15.13) is recorded during the impact phase (i.e., the phase during which the subject lands over the force platform at the end of the fly phase). During this phase, the protective behavior to safeguard the injured limb is evident. If the injured limb has not yet fully reached its functional and proprioceptive capacities, an important difference can be seen in this phase, between the limbs, in the force signal recorded by the force platform. The difference between values (injured limb versus healthy limb) must be less than 10%.

For further details concerning the SPT protocol and data interpretation, please see the dedicated article (Bisciotti 2006; Bisciotti et al. 2011).

<1%; sample rate 1000 Hz; sample depth 140 bit) record the force signal during the jump movement.

15.2.3 Isokinetic Tests

The isokinetic tests may be performed in two modalities: concentric modality and eccentric modality.

15.2.3.1 Concentric Modality

The test must be done with the non-injured leg first. The suggested velocities to be used during the test are a slow velocity (between 30 and 40°/s) and a high velocity (between 300 and 400 °/s). The slow velocity is useful for recording the peak force. Indeed, being the force-velocity relationship valuable also in isokinetic condition, the highest force values are achieved at low speeds (Bisciotti 2000). The higher speeds are used to study the mechanical behavior of the muscle at high speed of contraction. Indeed, it is important to remember that the muscle behavior is "speed-dependent," in

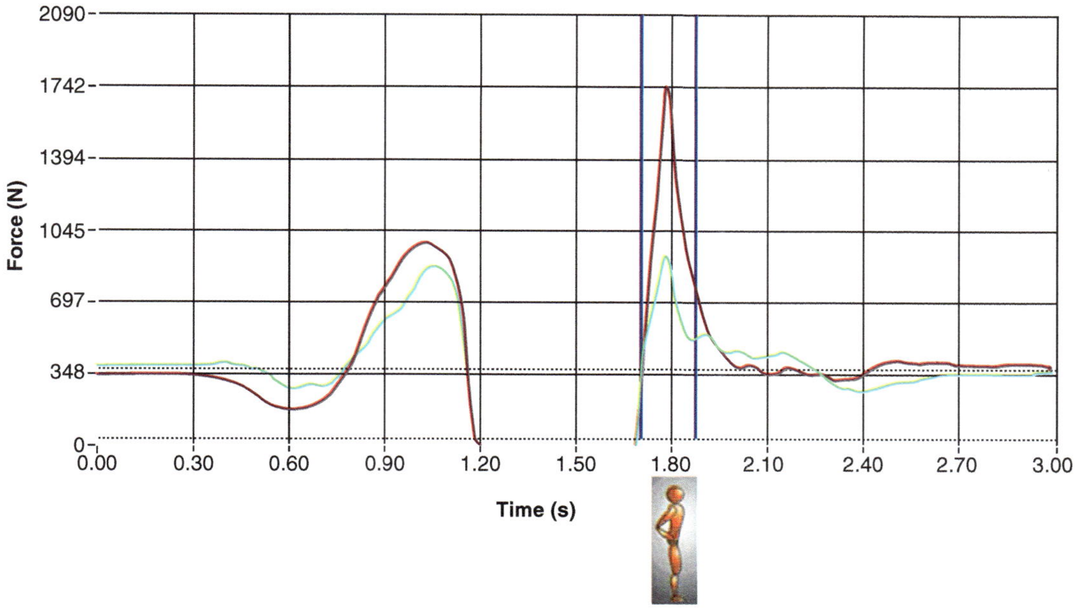

Fig. 15.13 The maximum impact force value during the landing phase (MFimp) highlights the protective mechanisms used by the subject. Red line: uninjured limb; green line: injured limb

other words, the muscle biomechanical behavior changes in relationship to the speed contraction (Devrome and MacIntosh 2018).

15.2.3.2 Eccentric Modality

The test must be done with the non-injured leg first. The eccentric test is performed using the continuous passive movement (CPM) modality. The suggested velocity is 60 °/s, because it is a good compromise between a good condition for the subject to be able to express a high degree of eccentric force and a controllable speed of movement. Since virtually all indirect muscle injuries occur during an eccentric contraction, the eccentric test is of paramount importance. Unfortunately, not all isokinetic machines have a CPM modality, and this represents a big limitation.

In the case of a proximal injury, the isokinetic test must be performed with extended leg as shown in Fig. 15.12. In the absence of the specific device shown in Fig. 15.14, an evaluation of the eccentric force (even if not isokinetic) can be made with the use of a hand dynamometer as shown in Fig. 15.15.

For isokinetic tests performed in both concentric and eccentric modality, the reference values are:

1. No pain on VAS scale during the test.
2. ii. All the calculated dynamometric values of the injured leg differ from the pre-injury values or from the healthy limb values by a maximum of 10%.

15.3 Hamstring RTT Laboratory Tests

The specific laboratory tests for RTT following hamstring injury are:

1. Isometric tests (Bisciotti 1998; Delvaux et al. 2014).
2. Isotonic tests (Croisier et al. 2002; Bisciotti 2003).
3. Isokinetic test (Menzel et al. 2013; Sanfilippo et al. 2013; Bisciotti et al. 2016).

Fig. 15.14 Isokinetic test performed with the use of a specific device (Kineo Globus, Italy) in the case of quadriceps proximal injury

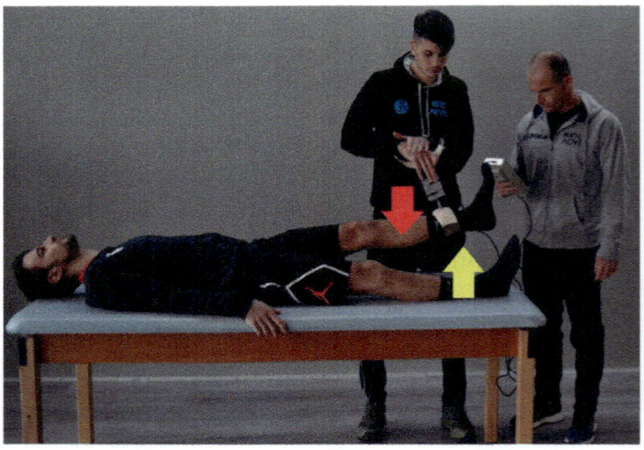

Fig. 15.15 Evaluation of the eccentric force performed with the use of a hand dynamometer in the case of proximal quadriceps injury. The operator makes a supramaximal push (red arrow) against the resistance of the subject (yellow arrow). Since the force exerted by the operator is greater than the resistance offered by the subject, this latter will be forced to perform an eccentric contraction

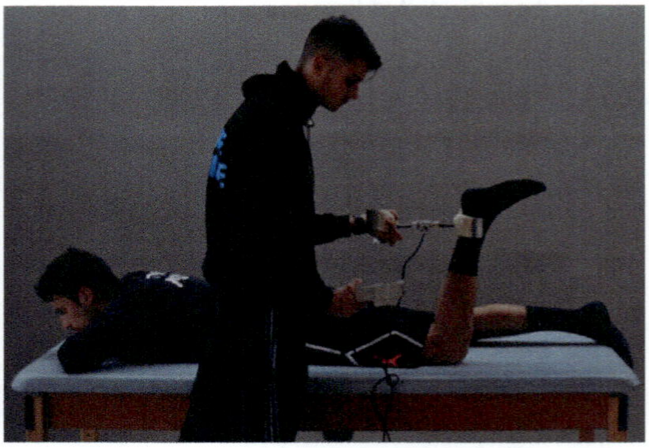

Fig. 15.16 Isometric test performed using a handle dynamometer in the case of middle third or distal hamstring injury

15.3.1 Isometric Test

Isometric test may be performed with a handle dynamometer (Fig. 15.16), with a normal leg curl machine equipped with a load cell (Fig. 15.17), or with a specific device (Kineo Globus, Italy) in isometric modality (Fig. 15.18). Since in isometric modality the force depends upon the joint angle, three different knee flexion angles (45 °, −90 °, −115 °) are considered.

In the case of proximal injury, the isometric test must be performed with the leg in extended position with a handle dynamometer (Fig. 15.19), or a load cell (Fig. 15.20), or a specific device (Kineo Globus, Italy) in isometric modality (Fig. 15.21). All tests must be done with the non-injured leg first. The reference values for isometric tests are:

1. No pain on VAS scale during the test.
2. All the calculated dynamometric values of the injured leg differ from the pre-injury values or from the healthy limb values by a maximum of 10%.

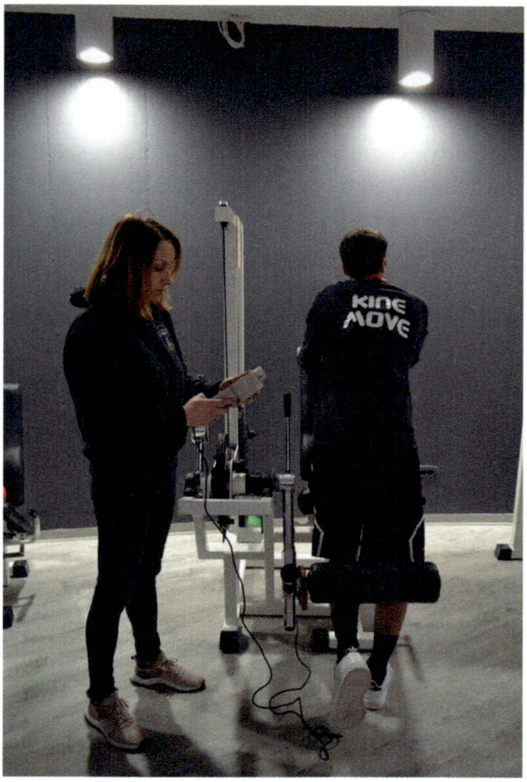

Fig. 15.17 Isometric test performed with the use of a leg curl machine equipped with a load cell in the case of middle third or distal hamstring injury

Fig. 15.18 Isometric test performed with the use of a specific device (Kineo Globus, Italy) in the case of middle third or distal hamstring injury

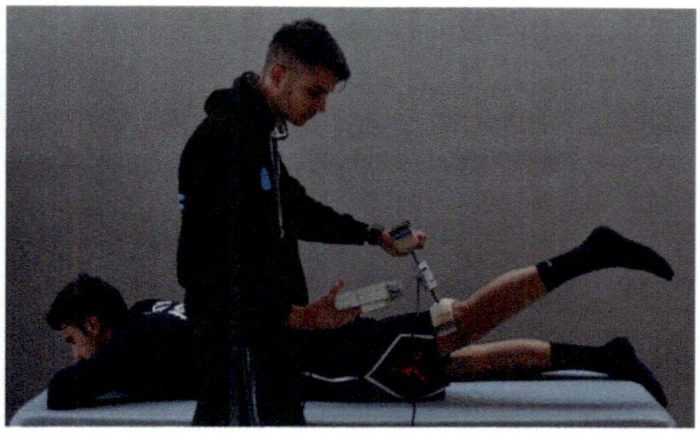

Fig. 15.19 Isometric test performed with the use of a handle dynamometer in the case of proximal hamstring injury

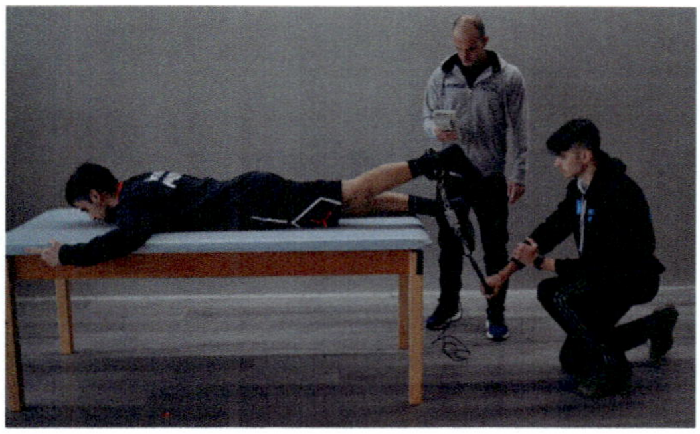

Fig. 15.20 Isometric test performed with the use of a load cell in the case of proximal hamstring injury

Fig. 15.21 Isometric test performed with the use of a specific device (Kineo Globus, Italy) in isometric modality in the case of proximal hamstring injury

15.3.2 Isotonic Test

Isotonic tests may be performed with a specific device (Kineo Globus, Italy) in isotonic modality. The execution of the test is different in case of middle third or distal injury (Fig. 15.22) or proximal injury (Fig. 15.23).

For the isotonic tests the parameters to be taken into consideration are the maximum force value and the peak of power production (Fig. 15.8) (Driss et al. 1998; Jaskólska et al. 1999).

All tests must be done with the non-injured leg first. The reference values are:

1. No pain on VAS scale during the test.
2. All the calculated dynamometric values of the injured leg differ from the pre-injury values or from the healthy limb values by a maximum of 10%.

Fig. 15.22 Isotonic test performed with the use of a specific device (Kineo Globus, Italy) in the case of hamstring middle third or distal injury

Fig. 15.23 Isotonic test performed with the use of a specific device (Kineo Globus, Italy) in the case of hamstring proximal injury

15.3.3 Isokinetic Tests

The isokinetic tests may be performed in two modalities: concentric modality and eccentric modality. The execution speeds are the same as those indicated for quadriceps isokinetic tests.

In the case of hamstring proximal injury, the isokinetic test must be performed with extended leg as shown in Fig. 15.24. In the absence of the specific device shown in Fig. 15.24, an evaluation of the eccentric force (even if not isokinetic) can be made with the use of a hand dynamometer as shown in Fig. 15.25.

The test must be done with the non-injured leg first.

For both concentric or eccentric modalities, the reference values are:

Fig. 15.24 Isokinetic test performed with the use of a specific device (Kineo Globus, Italy) in the case of hamstring proximal injury

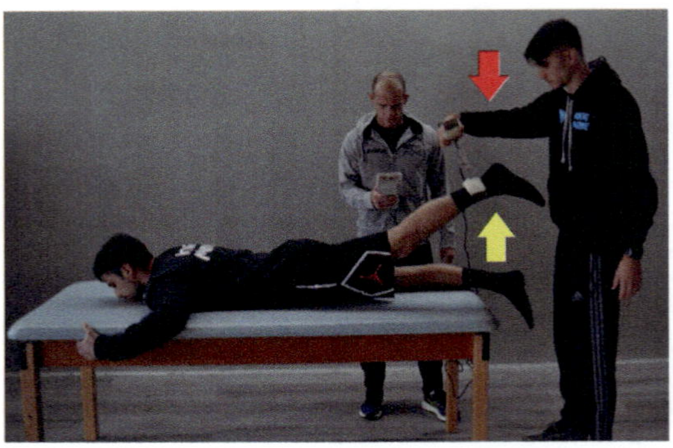

Fig. 15.25 Evaluation of the eccentric force performed with the use of a hand dynamometer in the case of proximal hamstring injury. The operator makes a supramaximal push (red arrow) against the resistance of the subject (yellow arrow). Since the force exerted by the operator is greater than the resistance offered by the subject, this latter will be forced to perform an eccentric contraction

1. No pain on VAS scale during the test.
2. All the calculated dynamometric values of the injured leg differ from the pre-injury values or from the healthy limb values by a maximum of 10%.

15.4 Adductor RTT Laboratory Tests

The specific laboratory tests for RTT following adductors injury are:

1. Isometric tests (Bisciotti 1998; Delvaux et al. 2014).
2. Isotonic tests (Croisier et al. 2002; Bisciotti 2003).
3. Isokinetic test (Menzel et al. 2013; Sanfilippo et al. 2013; Bisciotti et al. 2016).

15.4.1 Isometric Tests

The isometric tests may be performed with the use of a handle dynamometer (Fig. 15.26) or with the use of a specific device (Kineo Globus, Italy) in isometric modality (Fig. 15.27).

The test must be done with the non-injured leg first. The reference values are:

1. No pain on VAS scale during the test.
2. The dynamometric values of the injured leg differ from the pre-injury values or from the healthy limb values by a maximum of 10%.

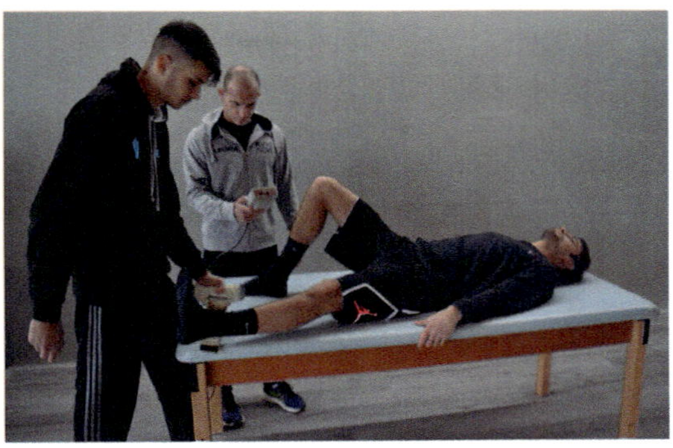

Fig. 15.26 Evaluation of the adductor isometric force performed with the use of a hand dynamometer

Fig. 15.27 Evaluation of the adductor muscle isometric force performed with the use of a specific device (Kineo Globus, Italy) in isometric modality

Fig. 15.28 Evaluation of the adductor muscle force performed with the use of a specific device (Kineo Globus, Italy) in isotonic modality

15.4.2 Isotonic Test

The isotonic test may be performed with the use a specific device (Kineo Globus, Italy) in isotonic modality (Fig. 15.28).As in the case of quadriceps and hamstring, also for adductor muscle isotonic tests, the parameters to be taken into consideration are the maximum force value and the peak of power production (Fig. 15.8) (Driss et al. 1998; Jaskólska et al. 1999).

The test must be performed with the non-injured leg first. The reference values are:

1. No pain on VAS scale during the test.
2. All the calculated dynamometric values of the injured leg differ from the pre-injury values or from the healthy limb values by a maximum of 10%.

15.4.3 Isokinetic Tests

Since the classic isokinetic machine cannot test the adductor muscle, it is necessary the use of a specific device (Kineo Globus, Italy) as shown in Fig. 15.29. In addition, it is possible to perform the isokinetic test both in concentric and eccentric modality.

In the absence of the specific device shown in Fig. 15.29, an evaluation of the eccentric (non-isokinetic) force can be made with the use of a hand dynamometer as shown in Fig. 15.30.

Fig. 15.29 Isokinetic evaluation of adductor muscles with the use of a specific device (Kineo Globus, Italy)

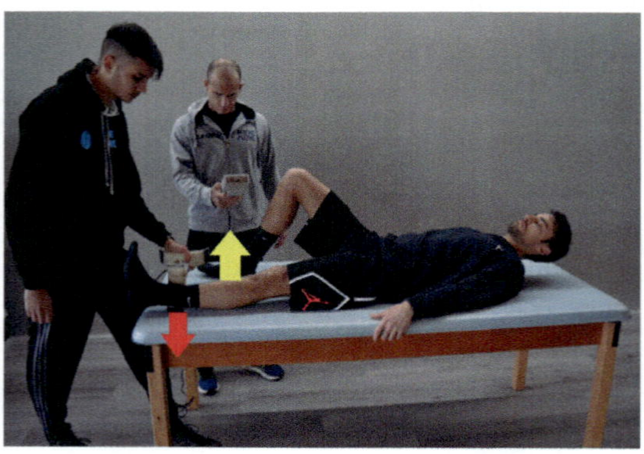

Fig. 15.30 Evaluation of the eccentric force performed with the use of a hand dynamometer in the case of adductor injury. The operator makes a supramaximal push (red arrow) against the resistance of the subject (yellow arrow). Since the force exerted by the operator is greater than the resistance offered by the subject, this latter will be forced to perform an eccentric contraction

The test must be done with the non-injured leg first.

Both for isokinetic tests performed in concentric and for those in eccentric modality, the reference values are:

1. No pain on VAS scale during the test.
2. All the calculated dynamometric values of the injured leg differ from the pre-injury values or from the healthy limb values by a maximum of 10%.

15.5 Soleus-Gastrocnemius RTT Laboratory Tests

The specific laboratory tests for RTT following soleus-gastrocnemius injury are:

1. Isometric tests (Bisciotti 1998; Delvaux et al. 2014).
2. Isotonic tests (Croisier et al. 2002; Bisciotti 2003).

Since the ROM of the ankle joint is very limited, the isokinetic tests are not recommended.

15.5.1 Isometric Tests

The isometric test may be performed both with a handle dynamometer (Fig. 15.29) or a calf machine equipped with a load cell (Fig. 15.30).

The test must be done with the non-injured leg first.

The reference values are:

1. No pain on VAS scale during the test.
2. The dynamometric values of the injured leg differ from the pre-injury values or from the healthy limb values by a maximum of 10% (Figs. 15.31 and 15.32).

15.5.2 Isotonic Test

The tests suggested for the soleus-gastrocnemius isotonic evaluation are:

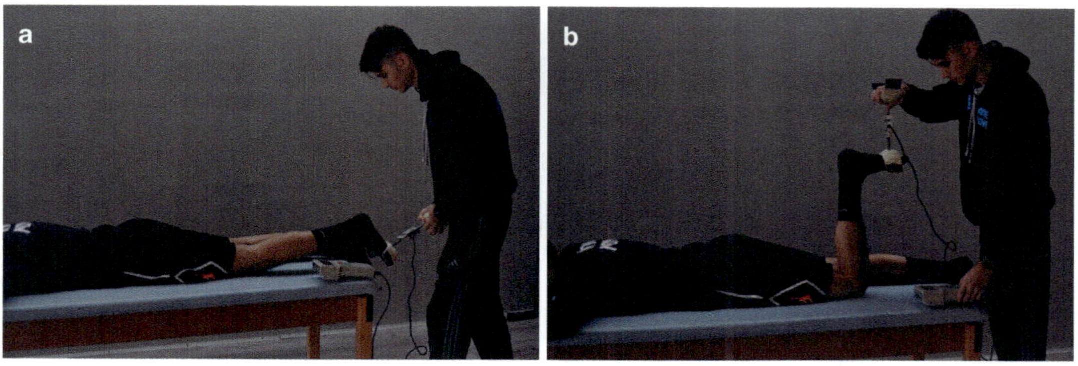

Fig. 15.31 Isometric evaluation of soleus-gastrocnemius muscles using a handle dynamometer. In the case of gastrocnemius evaluation, the knee joint must be kept extended (**a**), while in the case of soleus evaluation, the knee joint should be kept slightly flexed (**b**)

Fig. 15.32 Isometric evaluation of gastrocnemius muscle using a standing calf machine equipped with a load cell (**a**) and isometric evaluation of soleus muscle using a sitting calf machine equipped with a load cell (**b**)

1. The synchro plates test specific for calves muscle (Bisciotti et al. 2016) as shown in Fig. 15.33. The shape of the signal recorded during the test is shown in Fig. 15.34.
2. The drop test performed with synchro plates (Bisciotti et al. 2016) as shown in Fig. 15.33. The shape of the signal recorded during the test is shown in Fig. 15.34.

Fig. 15.33 The synchro plates test specific for soleus-gastrocnemius muscles

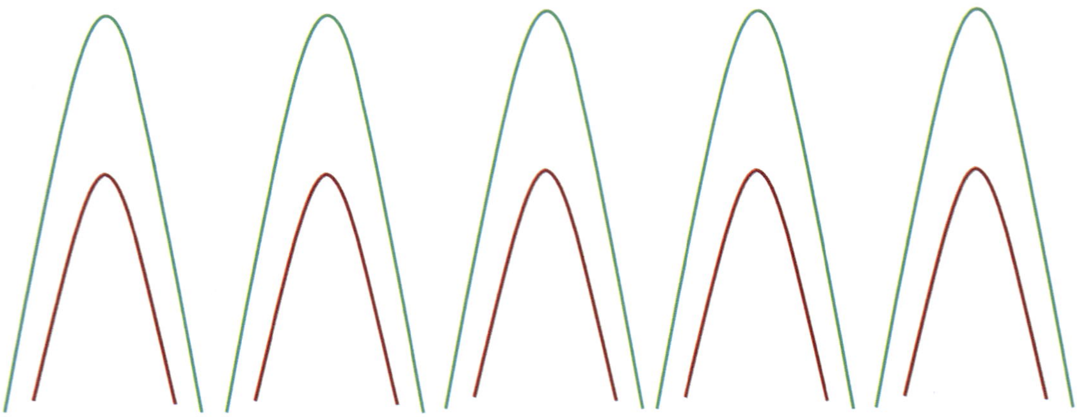

Fig. 15.34 The signal recorded during the synchro plates test specific for soleus-gastrocnemius muscles. In red the signal of the injured led and in green the signal of the non-injured leg

15.5.3 Synchro Plates Test Specific for Calves Muscles

During the synchro plates test specific for calves muscles, the subject must perform a series of six to ten rebound jumps performed with the knee in maximum extension.

The reference values are:

1. No pain on VAS scale during the test.
2. The dynamometric values of the injured leg differ from the pre-injury values or from the healthy limb values by a maximum of 10%.

15.5.4 Drop Test Performed with Synchro Plates

During drop test (Fig. 15.35) performed with synchro plates, the subject must perform a drop jump falling from a height between 30 and 40 centimeters.

The reference values are:

1. No pain on VAS scale during the test.
2. The dynamometric values of the injured leg differ from the pre-injury values or from the healthy limb values by a maximum of 10% (Fig 15.36).

Fig. 15.35 The drop test performed with synchro plates

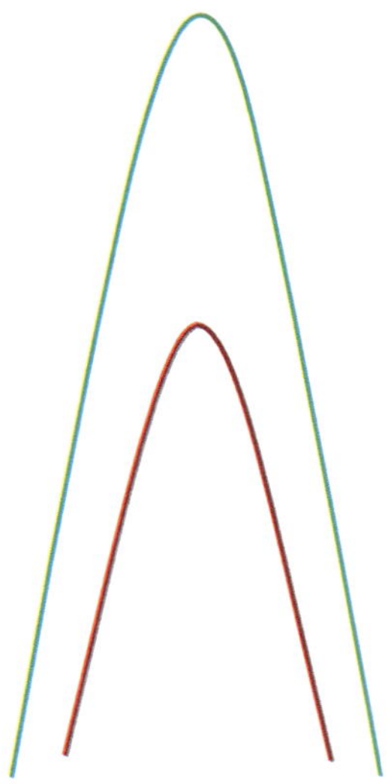

Fig. 15.36 The signal recorded during the drop jump performed with synchro plates. In red the signal of the injured leg and in green the signal of the non-injured leg

15.6 Short External Hip Rotator Muscles RTT Laboratory Tests

The specific laboratory test for RTT following external hip rotator muscles injury is simply an isometric test:

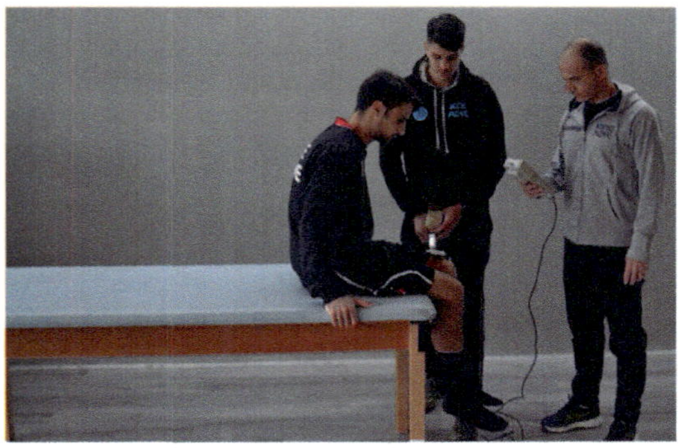

Fig. 15.37 Pace and Nagle test performed with the use of dynamometer

15.6.1 Isometric Tests

The specific assessment for short external hip rotator muscles injury is based on the Pace and Nagle maneuver tests (Delp et al. 1999).Pace and Nagle Maneuver Tests

The patient is in sitting position, and the operator places the hand dynamometer at knee level requesting to the athlete to perform a resisted hip abduction (Fig. 15.37). The test must be done with the non-injured leg first.The reference values are:

1. No pain on VAS scale during the test.
2. The dynamometric values of the injured leg differ from the pre-injury values or from the healthy limb values by a maximum of 10%.

References

Bisciotti GN, Quaglia A, Belli A, et al. Return to sports after ACL reconstruction: a new functional test protocol. Muscles Ligaments Tendons J. 2016;06:499–509.

Bisciotti GN, Sannicandro I, Manno R. Cambiamenti nel pattern di attivazione muscolare balistico a seguito di due diverse tecniche ricostruttive di legamento crociato anteriore del ginocchio. Med Sport. 2011;64(2):185–200.

Bisciotti GN, Volpi P, Alberti G, Aprato A, Artina M. Auci A et al. BMJ Open Sport Exerc Med. 2019 Oct 15;5(1):e000505. https://doi.org/10.1136/bmjsem-2018-000505.

Bisciotti GN. I valori di forza isometrica possono costituire un valore predittivo della performance dinamica? Coach Sport Sci J. 1998;3(1):47–58.

Bisciotti GN. La validazione di una nuova batteria di test per la quantificazione dello squilibrio muscolare: il Tesys Globus Evaluation System. New Athlet Res Sci Sport. 2006;198:5–15.

Bisciotti GN. Per ritrovare la funzionalità. Sport e Medicina. 2000;6:43–7.

Bisciotti GN. Valutazione come puzzle. Sport Medicina. 2003;2:51–8.

Chandramouli K, Theuerkauf P. Effect of knee angle on quadriceps strength and activation after anterior cruciate ligament reconstruction. J Appl Physiol. 2015 Aug 1;119(3):223–31.

Croisier J-L, Forthomme B, Namurois M-H, et al. Hamstring muscle strain recurrence and strength performance disorders. Am J Sports Med. 2002;30: 199–203.

Delp SL, Hess WE, Hungerford DS, Jones LC. Variation of rotation moment arms with hip flexion. J Biomech. 1999;32(5):493–501.

Delvaux F, Rochcongar P, Bruyere O, et al. Return-to-play criteria after hamstring injury: actual medicine practice in professional soccer teams. J Sports Sci Med. 2014;13:721–3.

Devrome AN, MacIntosh BR. Force-velocity relationship during isometric and isotonic fatiguing contractions. J Appl Physiol. 2018;125(3):706–14.

Driss T, Vandewalle H, Monod H. Maximal power and force-velocity relationship during cycling and cranck-ing exercises in volleyball players. Correlation with the vertical jump test. J Sports Med Phys Fitness. 1998;4(38):286–93.

Jaskólska A, Goossens P, Veenstra B, Jaskólski A, Skinner JS. Comparison of treadmill and cycle ergometer measurements of force-velocity relationship and power output. Int. J. Sports Med. 1999;20:192–7.

Knapik JJ, Wright JE, Mawdsley RH, Braun J. Isometric, isotonic, and isokinetic torque variations in four muscle groups through a range of joint motion. Phys Ther. 1983 Jun;63(6):938–47.

Menzel H-J, Chagas MH, Szmuchrowski LA, et al. Analysis of lower limb asymmetries by isokinetic and vertical jump tests in soccer players. J Strength Cond Res. 2013;27:1370–7.

Sanfilippo JL, Silder A, Sherry MA, et al. Hamstring strength and morphology progression after return to sport from injury. Med Sci Sports Exerc. 2013;45:448–54.

Verchoshansky YV. Componenti e struttura dell'impegno esplosivo di forza. SdS. 1996;34:15–21.

Wilson GJ, Lyttle D, Ostrowski KJ, Murphy AJ. Assessing dynamic performance: a comparison of rate of force development tests. J Strength and Cond Assoc. 1995;9:176–81.

The Field Tests for RTT Decision-Making Process

16

16.1 Introduction

After a severe muscle injury, the alteration of movement pattern may be particularly evident during the execution of movements such as monopodalic jumps (single leg jumping), changes of direction (sidestep cutting), and high speed run (Bonacci et al. 2009; Heiderscheit et al. 2010). This alteration could remain for a long period depending of the injury severity (Eliasson et al. 2018).

Rehabilitation processes can be considered finished with anatomical healing for sedentary patients, but the same is not true for sport subjects. Indeed, in sport subjects, the resumption of full functionality is essential. This aspect becomes more and more important when the performance level of the subject grows (Bisciotti et al. 2016). For this reason, the RTT decision-making process must absolutely be based on a battery of tests giving reliable information about the functionality of the injured muscle group. This chapter will address in detail the field tests proposed by the "Italian consensus conference statement on return to play after lower limb muscle injury in football." For each test, the muscle group for which the test is suitable will be specified.

16.2 Braking Test

16.2.1 Protocol

The braking test protocol (Bisciotti et al. 2015; Bisciotti and Volpi 2018) provides a preliminary test consisting in a maximal sprint for 30 m (Fig. 16.1). Once the sprint is performed, the subject is asked to make another sprint over the same distance at 90% of the maximum speed recorded during the preliminary test and to stop where a skittle placed at a specific distance from the end of the 30 meter sprint. The protocol provides for three tests of 30 meters, the first of which has the stop-skittle placed at 8 m and the second and the third at 6 and 4 m from the end of the 30 m sprint, respectively. The purpose of the braking test is to quantify in an objective manner the effectiveness of the contraction of the flexor muscles during the braking phase. In fact, during the arrest time, the subject must dissipate the kinetic energy (C) taken during the sprint from a certain value (dependent on his mass and the speed reached) to zero. If the C value and the stopping distance (s) are known, the value of the negative power expressed by the subject during the braking phase can be calculated. It is, therefore, possible to indirectly calculate the effectiveness of the flexor muscles co-contraction through the following formula:

$$P(W) = (0.5M \cdot V2) / Tf \qquad (16.1)$$

G. N. Bisciotti et al., *Return-to-Play after Lower Limb Muscle Injury in Football*,
https://doi.org/10.1007/978-3-030-84950-4_16

Fig. 16.1 Braking test. The test is performed on three different braking distances (i.e., 8, 6, and 4 m). Braking test 8 m. Braking test 6 m. Braking test 4 m

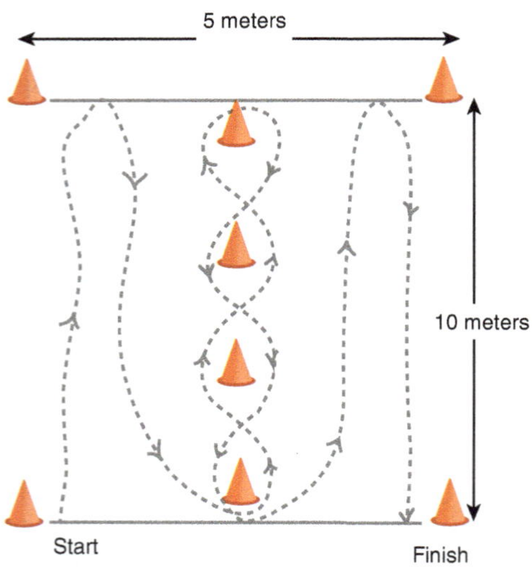

Fig. 16.2 Illinois agility test scheme. The test area has a length of 10 m and a width of 5 m

in which P is the negative power expressed in W, M is the subject's mass, V is the average speed.

reached during the sprint, and Tf is the braking time, the time taken to reach V0 (zero speed, the full stop).

Knowing that Tf is equal to:

$$V2.s / a \tag{16.2}$$

in which *s* is the braking and *a* the acceleration, and knowing that the *a* value is equal to:

$$V2 / 2.s \tag{16.3}$$

in which V2 is the square of the speed reached by the athlete during the sprint.

Replacing the *a* value derived from (16.3) in (16.2) and replacing in (16.1) the Tf value derived from (16.2), it is possible to calculate the P value, that is, the negative power expressed during the contraction of the flexor muscles into the braking phase. An electronic spreadsheet specifically conceived allowed the calculation. The minimum acceptable value for the test was fixed in 22 W/kg^{-1}.

16.2.2 Indication

The braking test is indicated for the hamstring assessment and, since the flexor muscles have a strong synergy with the extensor muscles, also for quadriceps injuries.

16.3 Illinois Agility Test

16.3.1 Protocol

In the Illinois Agility Test (Lockie et al. 2013; Hachana et al. 2014), the subject must perform at maximum speed the entire path according to the scheme shown in Fig. 16.2. The start is freely determined by the athlete. The chronometric result represents the final score of the test (Table 16.1).

Table 16.1 Illinois agility test reference values

Ranking	Lap results male (s)	Lap results female (s)
Excellent	<15.2	<17
Good	16.1–15.2	17–17.9
Medium	18.1–16.2	18.0–21.7
Insufficient	18.2–19.3	21.8–23.0
Very insufficient	>19.3	> 23

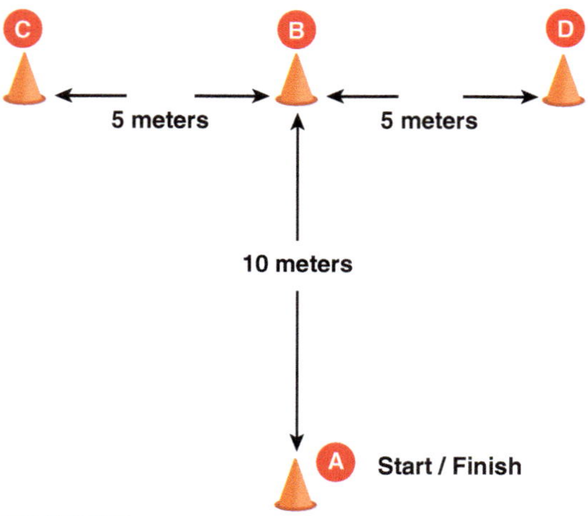

Fig. 16.3 Agility T-test scheme

16.3.2 Indication

Since the Illinois Agility Test reproduces the typical soccer performance model (i.e., acceleration, deceleration, and changes of direction), it is indicated for all muscle injuries. Furthermore, since the Illinois Agility Test is asymmetric (Bisciotti 2015; Hachana et al. 2013), we recommend execution in its modified format formulated by Rouissi et al. (2016).

16.4 Agility T-Test

The agility T-test (Sassi et al. 2009) is performed following the scheme shown in Fig. 16.3. The athlete starts at cone A and sprints to cone B and touches the base of the cone with the right hand. He or she then sprints left to cone C and touches the base of the cone with the left hand. He or she then sprints right to cone D and touches the base of the cone with the right hand. Finally, he or she sprints back to cone B and touches the base with the right hand before running backward to the start (cone A). The chronometric result represents the final score of the test (Table 16.2).

16.4.1 Indication

The agility test is suggested for adductor muscle, hip short external rotator, and iliopsoas injuries.

16.5 Carioca Test

16.5.1 Protocol

The carioca test (Kong et al. 2012; Jang et al. 2014) is performed by having the patient run laterally two lengths of 12 m (Fig. 16.4) with a crossover step. The patient runs the course from left to right and then from right to left until reaching the maximum speed. The reference criterion is the absence of pain on VAS scale.

16.5.2 Indication

The carioca test is indicated for adductor and hip short external rotator injuries.

16.6 Kicking Test

16.6.1 Protocol

The kicking test is performed following the scheme shown in Fig. 16.5. Five balls are placed in front of the soccer goal, and ball n° 3 is placed on the penalty spot. The athlete starts in sprint and kicks at the goal (at maximum power) every balls, returning each time (always in sprint) to the starting spot. The tested athlete must kick three times with the injured leg and twice with the non-injured leg. In this way, the injured leg is tested both as a kick leg and as a support leg. The reference value is absence of pain on VAS scale.

Table 16.2 Agility T-test reference values

Ranking	Lap results male (s)	Lap results female (s)
Excellent	<9.50	< 10.50
Good	9.51–10.50	10.51–11.50
Average	10.51–11.50	11.51–12.50
Poor	>11.50	> 12.50

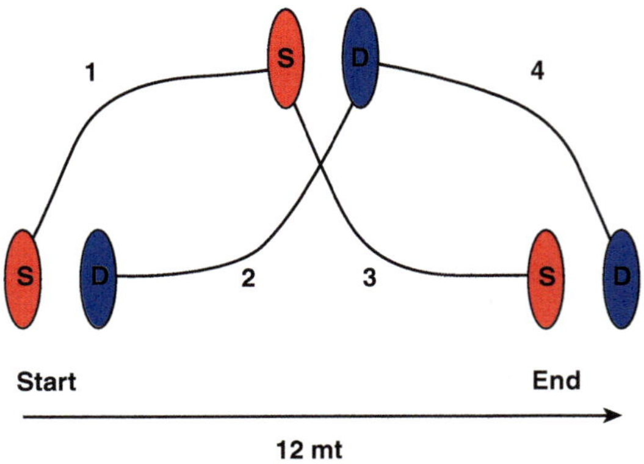

Fig. 16.4 Fig. 16.4: carioca test scheme

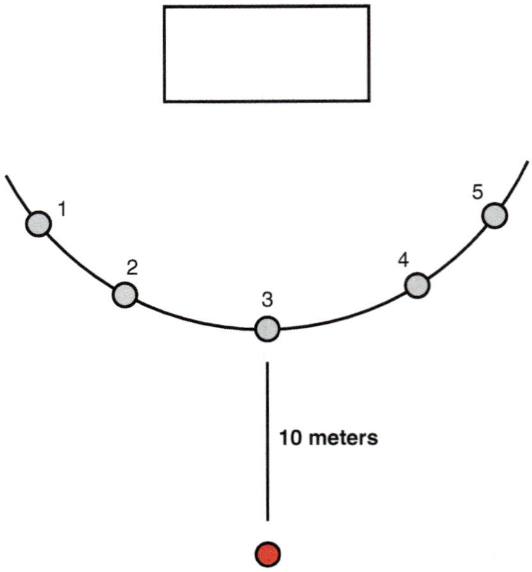

Fig. 16.5 Kicking test scheme

16.6.2 Indication

The kicking test is indicated for injuries to quadriceps, adductors, hip short external rotator muscles, and iliopsoas.

16.7 Retro-Run Test

16.7.1 Protocol

During the retro-run test (Brumitt et al. 2013), the subject must perform a backward running over the distance of 20 meters at maximum speed and with three different modalities (Fig. 16.6):

1. Minimum amplitude.
2. Medium amplitude.
3. Maximum amplitude.

The reference value is absence of pain on VAS scale.

16.7.2 Indication

The retro-run test is indicated for hamstring muscle injuries.

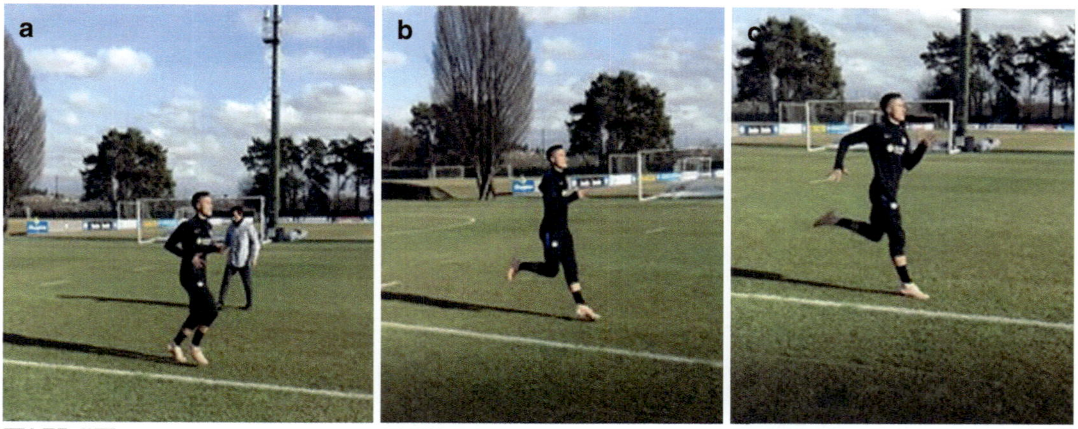

Fig. 16.6 The retro-run test performed following the three different modalities, that is, minimum amplitude (**a**), medium amplitude (**b**), and maximum amplitude (**c**). Retro-run test minimum amplitude. Retro-run test medium amplitude. Retro-run test maximum amplitude

References

Bisciotti GN, Quaglia A, Belli A, Carimati G, Volpi P. Return to sports after ACL reconstruction: a new functional test protocol. Muscles Ligaments Tendons J. 2016 Oct-Dec;6(4):499–509.

Bisciotti GN, Volpi P. Return to play. In: Volpi P, editor. Football doctor manual. Trento: Edra Edition; 2018. p. 247–59.

Bisciotti GN. Return to play after a muscle lesion. In: Volpi P, editor. Arthroscopy in sport. Cham: In, Springer Edition; 2015.

Bonacci J, Chapman A, Blanch P, Vicenzino B. Neuromuscular adaptations to training, injury and passive interventions: implications for running economy. Sports Med. 2009;39(11):903–21.

Brumitt J, Heiderscheit BC, Manske RC, et al. Lower extremity functional tests and risk of injury in division III collegiate athletes. Int J Sports Phys Ther. 2013;8:216–27.

Eliasson P, Agergaard AS, Couppé C, Svensson R, Hoeffner R, Warming S, Warming N, Holm C, Jensen MH, Krogsgaard M, Kjaer M, Magnusson SP. The Ruptured Achilles Tendon Elongates for 6 Months After Surgical Repair Regardless of Early or Late Weightbearing in Combination With Ankle Mobilization: A Randomized Clinical Trial. Am J Sports Med. 2018 Aug;46(10):2492–502.

Hachana Y, Chaabène H, Ben Rajeb G, et al. Validity and reliability of new agility test among elite and subelite under 14- soccer players. PLoS One. 2014;9(4):e95773.

Hachana Y, Chaabene H, Nabli MA, et al. Test-retest reliability, criterion-related validity, and minimal detectable change of the Illinois agility test in male team sport athletes. J Strength Cond Res. 2013;27: 2752–9.

Heiderscheit BC, Sherry MA, Silder A, Chumanov ES. Thelen DG.J Hamstring strain injuries: recommendations for diagnosis, rehabilitation, and injury prevention. Orthop Sports Phys Ther. 2010 Feb;40(2):67–81.

Jang SH, Kim JG, Ha JK, et al. Functional performance tests as indicators of returning to sports after anterior cruciate ligament reconstruction. Knee. 2014;21:95–101.

Kong DH, Yang SJ, Ha JK, et al. Validation of functional performance tests after anterior cruciate ligament reconstruction. Knee Surg Relat Res. 2012;24: 40–5.

Lockie RG, Schultz AB, Callaghan SJ, Jeffriess MD, Berry SP. Reliability and validity of a new test of change-of-direction speed for field-based sports: the change-of-direction and acceleration test (CODAT). J Sports Sci Med. 2013;12(1):88–96.

Rouissi M, Chtara M, Berriri A, et al. Asymmetry of the modified Illinois change of direction test impacts young elite soccer players' performance. Asian J Sports Med. 2016;7:e33598.

Sassi RH, Dardouri W, Yahmed MH, Gmada N, Mahfoudhi ME, Gharbi Z. Relative and absolute reliability of a modified agility T-test and its relationship with vertical jump and straight sprint. J Strength Cond Res. 2009 Sep;23(6):1644–51.

17 Case Report: Return to Play and Return to Training After Quadriceps Injury

17.1 Introduction

Quadriceps femoris (QF) indirect injuries are relatively frequent in sports such as football, rugby, and American football (Kary 2010). The etiopathogenesis of this type of injury is generally represented by a violent and sudden QF eccentric contraction during the phase of adjustment and control of the knee joint flexion. The muscle of the QF complex in which the greatest number of indirect trauma is recorded is the rectus femoris (RF) (Young et al. 1993; Hasselman et al. 1995; Hughes et al. 1995; Garrett 1996; Cross et al. 2004). The factors determining the greatest risk of injury for RF are essentially that RF is the only biarticular muscle of the group, rich in type II fibers, and it has a rather complex muscle-tendon architecture (Hughes et al. 1995; Hasselman et al. 1995; Järvinen et al. 2005). The biomechanically highest-risk movement is represented by a violent eccentric contraction performed with the knee joint flexed, simultaneous with an extension of the hip joint, like during the kicking movement (Garrett 1996). In football the QF injuries, and therefore in particular the RF injuries, involve recovery times ranging on average from 39 to 45 days, depending on whether these injuries are mid-distal or proximal, respectively (Balius et al. 2009). In football, 88% of QF lesions, as staged by MRI examination, occur at the RF level, and the risk of incurring this type of injury is much higher during competition than during normal training sessions (1.1 versus 0.3/1000 h). During the first half of the competition, 62% of all QF injuries are recorded, and the peak of the injury risk is observed between the 16th and 45th minute of the competition, a period during which about 40% of all injuries to the QF occur. About 28% of injuries occur during the act of shooting on goal. This is exactly the opposite situation with respect to the hamstring injuries, for which the risk during an action that involves shooting on goal is reduced to only 1.5%. During the sports season the QF injuries peak is recorded at the end of the summer preparation period, during which the movement of shooting on goal during training sessions is generally more frequent. After this period, the QF injuries does not record statistically significant increases during the season. Once again, this is the opposite of what has been observed for hamstring injuries. Indeed, hamstring injury risk constantly increases throughout the season, and its peak is in the last 2 months of training. The majority of QF injuries—the same is true for hamstring injuries—have a "non-contact" etiology (96%), and the reinjury rate is approximately equal to 13%. A team composed by 25 players must expect around three QF injuries per season, for a total absence of around 50 days. On average, a QF injury results in 18 days of absence, which in turn entails the loss of approximately 12 training sessions and three competitions. Nineteen percent of injuries can be classified as "severe" and result in

G. N. Bisciotti et al., *Return-to-Play after Lower Limb Muscle Injury in Football*,
https://doi.org/10.1007/978-3-030-84950-4_17

an absence of more than 4 weeks. The QF injuries do not show a significant statistical difference, concerning the days of absence from training and competition, depending on whether the injuries occur in competition or during training sessions (20 versus 17 days). This is, once again, the opposite situation with respect to hamstring injuries. Finally, to our knowledge in the literature, there are no descriptions of massive tears in the vastus intermedius muscle (VI).

17.2 Case Report

FF, a 23-year-old footballer playing in Lega Pro, came to our medical observation, following a massive injury to the left VI. The injury had occurred 10 days earlier during a training session, when he performed a movement and slipped on the supporting leg (i.e., the injured limb) thus elongating the QF in an abrupt and sudden manner. At the time of the clinical evaluation, the patient was walking with the help of crutches. The injured limb appeared swollen, turgid, and painful to the palpation. The slightest passive elongation generated a severe painful symptomatology. The MRI examination showed at the VI middle-proximal third level, an oval formation of 4.4 × 6.4 cm in axial diameter and 11.7 cm of longitudinal extension. The lesion area was characterized by inhomogeneity and hyperintensity of the signal in the T1-weighted and long-TR sequences. These images were consistent with massive hematoma in high-grade indirect injury outcomes. There was also an important edematous area both in the remaining portions of the VI and in the vastus lateral, vastus medialis, sartorius, and tensor fascia lata muscles. A thin fluid layer, with a likely hematic character, was also present in the intrafascial area at the level of all the muscles of the anterior compartment of the thigh. Furthermore, an imbibition of the subcutaneous adipose tissue was present (Figs. 17.1, 17.2, and 17.3). The injury was classified as third degree following the classification proposed by the Italian Consensus Conference on guidelines for conservative treatment on lower limb muscle injuries in athletes (Bisciotti et al. 2018). The athlete started his rehabilitation program with phase I.

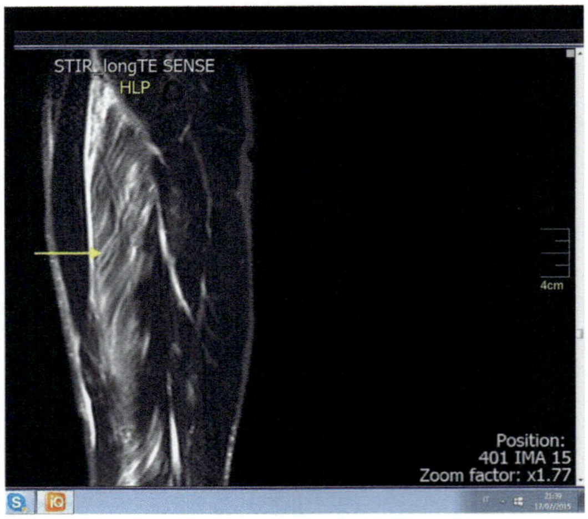

Fig. 17.1 Coronal STIR MR image showing the longitudinal extension of edema following the injury (arrow)

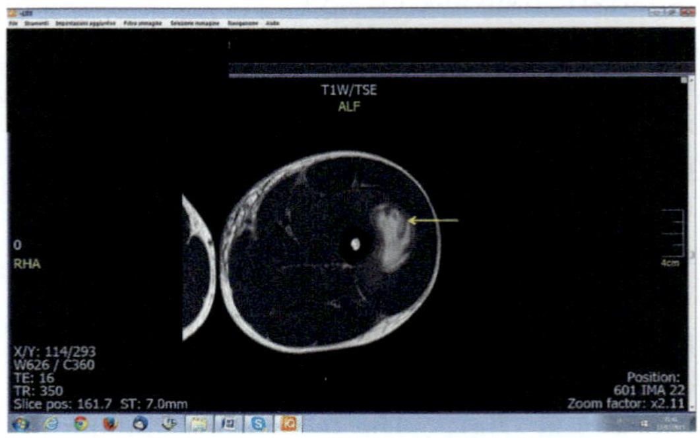

Fig. 17.2 T1-weighted axial MR image showing an oval formation of 4.4 × 6.4 cm in axial diameter and 11.7 cm in longitudinal extension, characterized by inhomogeneity and hyperintensity of the signal (arrow)

17.2.1 Phase I

Phase I, which lasted 15 days, was based on:

1. Progressive recovery of the ROM.
2. Gait recovery.
3. Mobilization of the injured leg.
4. Isometric contractions at progressive intensity both in open kinetic chain (OKC) and in closed kinetic chain (CKC).

In the first phase, the exclusive use of the isometric contractions is justified by the fact that the lesional gap (i.e., the interruption of the muscle fibers' continuity) cannot be solicited yet, neither in concentric contraction nor in eccentric contraction. Indeed, an excessive stress on the lesional gap can cause its further diastasis. The isometric contraction is important as it provides a tonic-functional stimulus, especially to the partially injured fibers, accelerating their regeneration processes. Furthermore, the isometric contraction maintains the muscle trophism, avoiding the risk of lesion gap aggravation. A graphic representation of the anatomo-pathological situation of this period is shown in Fig. 17.4.

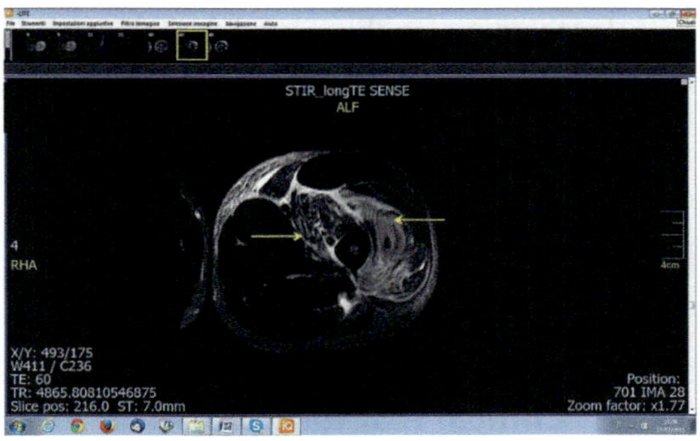

Fig. 17.3 Axial MR STIR image highlighting the important edematous area both in the remaining portions of the VI and in the vastus lateral, vastus medial, sartorius, and tensor fascia lata muscles (arrows)

17.2.2 Phase II

The transition criteria from the first to the second rehabilitation phase were based on the clinical assessment and the imaging examination (Bisciotti et al. 2018):

Clinical and functional criteria:

1. No pain during maximum isometric contraction.
2. No pain during passive stretching.
3. No pain during active stretching tests.
4. Full ROM of the hip and knee joints.

Imaging criteria:

1. Decrease of the lesion gap at MRI (or US) examination.
2. Presence of granulation repair tissue within the lesion gap in US examination.

Phase II, which lasted 30 days, was based on:

1. Concentric contractions with progressively increasing intensity both in OKC and in CKC.
2. Active and passive stretching.

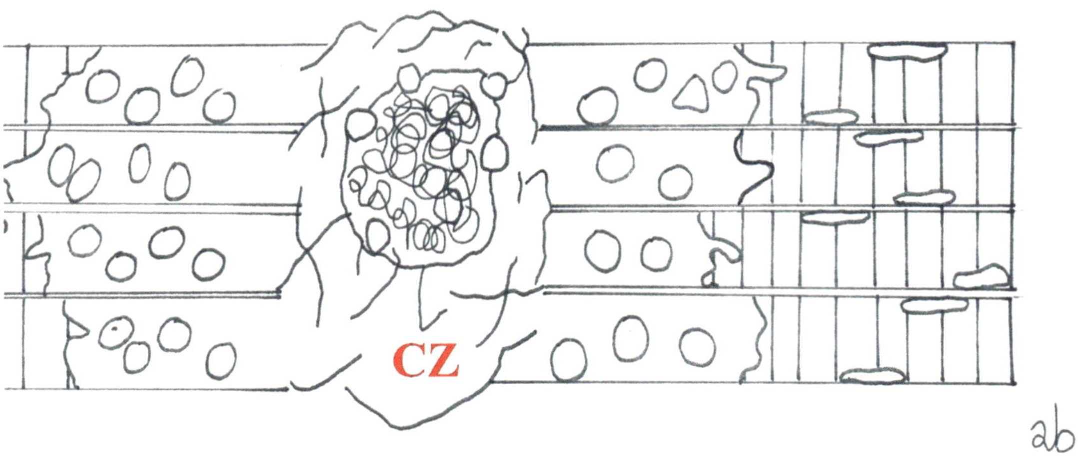

Fig. 17.4 In this phase the necrotized parts of the muscle fibers are removed by the macrophages while, at the same time, the formation of the scarring connective tissue within the central zone (CZ i.e., the area of the lesion gap) has begun

3. Introduction of straight run with progressive increase of speed.
4. Proprioceptive training.

Normally, in this phase, the formation of granulation tissue within the CZ starts to be observed at US examination. However, an eccentric stimulus would still be potentially able to diastase the lesion gap. For this reason, during this phase, the muscle can only be stimulated in concentric modality so as to favor the rapprochement of the injured muscle stumps, avoiding the risk of lesional gap aggravation. A graphic representation of the anatomo-pathological situation of this period is shown in Fig. 17.5.

17.2.3 Phase III

The transition criteria from the second to the third rehabilitation phase were based on the clinical assessment and the imaging examination (Bisciotti et al. 2018):

Clinical and functional criteria:

1. No pain during maximal concentric contractions.
2. No pain during submaximal eccentric contractions.

Imaging criteria:

1. Disappearance of the lesion gap in MRI (or US) examination.
2. Presence of compact granulation repair tissue within the lesion gap in US examination.

Fig. 17.5 During this period, the scar zone in the CZ area has further condensed and decreased in size, and the repaired myofibers fill the residual gap of the CZ area

Phase III, which lasted 27 days, was based on:

1. Eccentric exercises with progressively increased intensity.
2. Run with changes of direction moments at progressive speed.
3. Football individual technique.

At this stage, the granulation tissue within the lesional gap begins to have a certain structural compactness (Fig. 17.6). The eccentric stimulation allows the correct orientation of the fibers in formation, avoiding an "anarchic" proliferation mode. Indeed, an "anarchic" proliferation would result in the formation of an excessive fibrotic area. Furthermore, in this period the following is crucial:

1. Conditioning the neo-fibers for eccentric contraction.
2. Restoring full mechanical functionality toward the eccentric stimulus.

17.3 Clinical Test for Return to Training

The return to training (RTT) decision-making process was based on the guidelines of the Italian Consensus Statement on return to play after lower limb muscle injury in football (Bisciotti et al. 2019). The following steps in clinical evaluation were respected:

1. Absence of clinical symptoms.
2. Absence of pain or tenderness during muscle palpation.
3. Absence of pain on passive and active stretching.
4. Absence of pain on isometric, concentric, and eccentric contraction.
5. Completion of the prescribed rehabilitation program.
6. MRI imaging assessment (Fig. 17.7).
7. Subjective feelings of the player were taken in account.

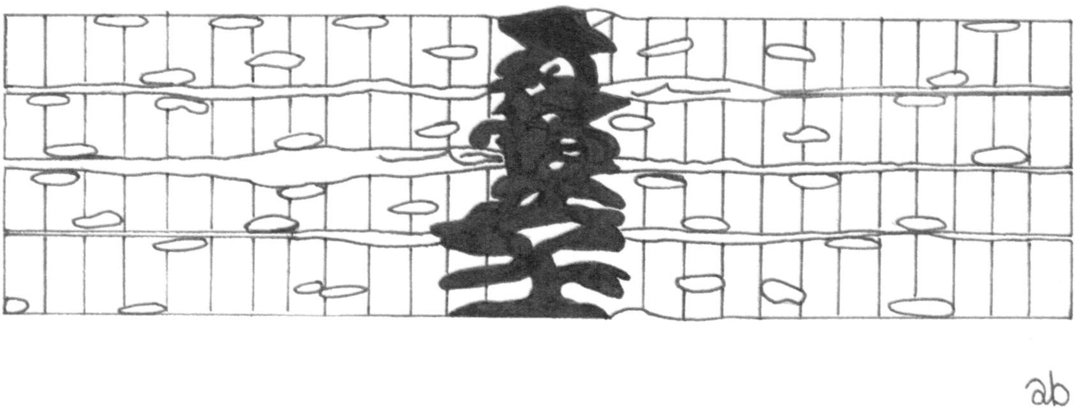

Fig. 17.6 During phase III, the scar area in the CZ area has further condensed and decreased in size, and the repaired myofibers fill the residual gap in the CZ area

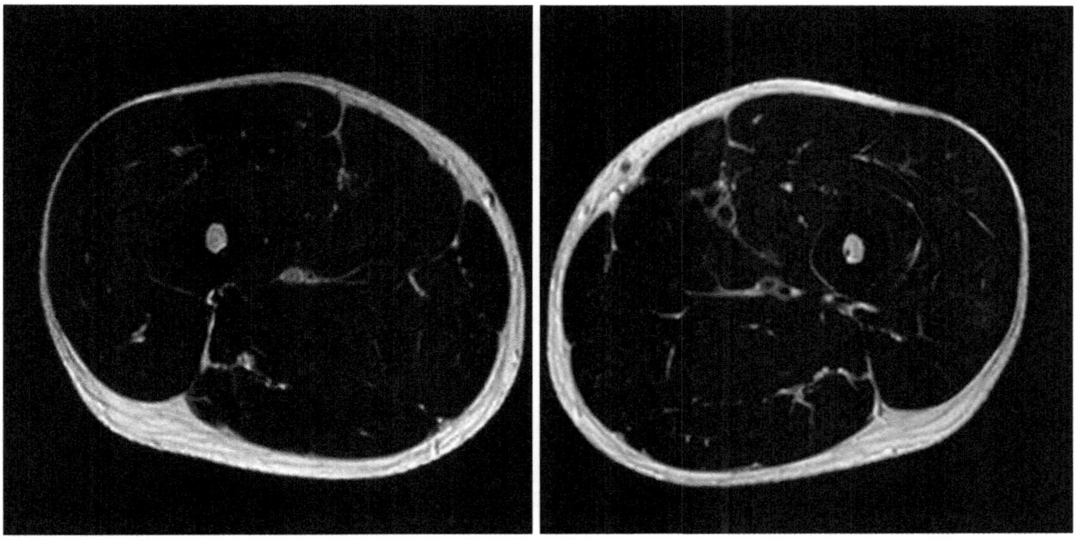

Fig. 17.7 Axial MRI demonstrating complete anatomical healing of the injured area

17.3.1 Specific Assessment

The specific assessment for RTT was based on the passive quadriceps stretch test (Witvrouw et al. 2003; Bouvier et al. 2017).

17.4 Laboratory Tests for RTT

The following laboratory tests were performed:

1. Quadriceps muscle strength was assessed by dynamometric tests. The values of dynamometric tests were > 90% of contralateral values (Bisciotti 2015).
2. Synchro plates test (Bisciotti et al. 2016). The values were > 90% of contralateral values (Bisciotti 2015).

17.5 Field Tests for RTT

The following field tests were performed:

1. Illinois Agility Test (Hachana et al. 2013; Raya et al. 2013; Bisciotti 2015; Negra et al. 2017; Bisciotti and Volpi 2018).
2. Braking test (Bisciotti 2015; Bisciotti and Volpi 2018).
3. Kicking test (Bisciotti 2015; Bisciotti and Volpi 2018).

17.6 Return to Play Tests

The player received clearance for the RTT, so he started the return to play (RTP) decision-making process. The process was based on GPS data acquisition (Bisciotti et al., 2019). The process lasted 30 days during which the following parameters were recorded:

1. Quantitative evaluation.
2. Qualitative evaluation.
3. Parameter analysis.

For an in-depth description of the parameters and categories, please see Chap. 4.

After 25 days, the clearance for the RTP was obtained. About 10 months after the injury, no reinjury occurred, and the subject is fully satisfied with the rehabilitation path.

17.7 Discussion

The VI injury is a rare event; to our knowledge in the literature, there are no descriptions of such lesion. The rarity of the VI lesion can be explained by the vastus intermedius being, from an anatomotopographic point of view, a monoarticular, deep muscle. Beyond the rarity of the case described, we would like to underline that the program was based on a precise therapeutic pathway. Consequently, the working methodology was kept as close as possible to the progress of the injury's biological repair. In particular, in each phase, the functional stress on the muscle was tailored to the existing possibility of response and functional adaptation of the biological tissue. Therefore, the therapeutic path was not dictated by a theoretical time limit, but rather linked to the biological repair processes of the muscle tissue.

17.8 Conclusions

Massive VI injuries are rare in football; nevertheless it can be the cause of a major time loss injury for the player. The rehabilitation must necessarily take into consideration the biological repair times of the muscle tissue to be treated. For this reason, imaging examinations (MRI and US) are very important monitoring tools for an objective decision on the passage fort the different steps of the rehabilitation phases.

References

Balius R, Maestro A, Pedret C. Central aponeurosis tear of the rectus femoris: practical sonographic prognosis. B J Sport Med. 2009;43(11):818–24.

Bisciotti GN. Return to play after a muscle lesion. In: Volpi P, editor. Arthroscopy in sport. Cham: Springer Edition; 2015.

Bisciotti GN, Quaglia A, Belli A, et al. Return to sports after ACL reconstruction: a new functional test protocol. Muscles Ligaments Tendons J. 2016;06:499–509.

Bisciotti GN, Volpi P. Return to play. In: Volpi P, editor. Football doctor manual. Trento: Edra Edition; 2018. p. 247–59.

Bisciotti GN, Volpi P, Amato M, et al. Italian consensus conference on guidelines for conservative treatment on lower limb muscle injuries in athlete. BMJ Open Sport Exerc Med. 2018;0:e000323. https://doi.org/10.1136/bmjsem-2017-000323.

Bisciotti GN, Volpi P, Alberti G, Aprato A, Artina M, et al. Italian consensus statement (2020) on return to play after lower limb muscle injury in football (soccer). BMJ Open Sport Exerc Med. 2019;15;5(1):e000505.

Bouvier T, Opplert J, Cometti C, et al. Acute effects of static stretching on muscle-tendon mechanics of quadriceps and plantar flexor muscles. Eur J Appl Physiol. 2017;117:1309–15.

Cross TM, Gibbs N, Houang MT, et al. Acute quadriceps muscle strains: magnetic resonance imaging features and prognosis. Am J Sports Med. 2004;32:710–9.

Garrett WE. Muscle strain injuries. Am J Sports Med. 1996;24:S2–8.

Hachana Y, Chaabene H, Nabli MA, et al. Test-retest reliability, criterion-related validity, and minimal detectable change of the Illinois agility test in male team sport athletes. J Strength Cond Res. 2013;27:2752–9.

Hasselman CT, Best TM, Hughes C, et al. An explanation for various rectus femoris strain injuries using previously undescribed muscle architecture. Am J Sports Med. 1995;23:493–9.

Hughes C, Hasselman CT, Best TM, et al. Incomplete, intrasubstance strain injuries of the rectus femoris muscle. Am J Sports Med. 1995;23:500–6.

Järvinen TAH, Järvinen TLN, Kääriäinen M, et al. Muscle injuries: biology and treatment. Am J Sports Med. 2005;33:745–64.

Kary JM. Diagnosis and management of quadriceps strains and contusions. Curr Rev Musculoskelet Med. 2010 Jul 30;3(1–4):26–31.

Negra Y, Chaabene H, Hammami M, et al. Agility in young athletes: is it a different ability from speed and power? J Strength Cond Res. 2017;31:727–35.

Raya MA, Gailey RS, Gaunaurd IA, et al. Comparison of three agility tests with male servicemembers: Edgren side step test, t-test, and Illinois agility test. J Rehabil Res Dev. 2013;50:951–60.

Witvrouw E, Danneels L, Asselman P, et al. Muscle flexibility as a risk factor for developing muscle injuries in male professional soccer players. Am J Sports Med. 2003;31:41–6.

Young JL, Laskowski ER, Rock MG. Thigh injuries in athletes. Mayo Clin Proc. 1993;68:1099–106.

18 Case Report: Return to Play and Return to Training After Hamstring Injury

18.1 Introduction

Football (soccer) is classified among sports with a high traumatic risk. This is unequivocally confirmed by a great number of epidemiological studies. Indeed, the percentage of professional male players suffering an injury during a sporting season is between 65 and 91% (Waldén et al. 2005). From an epidemiological point of view, in the football player, the anatomical location recording the greatest number of injuries is the thigh, followed by the ankle joint, the pubic area, and the knee joint. The injuries of the thigh flexor muscles (classified in the Anglo-Saxon nomenclature with the collective term of hamstring) represent, among the muscle-tendon injuries, the most recurrent trauma, accounting for about 17% of all injuries. This type of injury is so frequent that a typical team of 25 players can statistically expect about seven hamstring injuries for each season (Ekstrand et al. 2011a, b). This results in approximately 2 weeks of training and competition lost for each injury event (Eirale and Ekstrand 2013). In terms of exposure per 1000 hours of play, the incidence of indirect hamstring injuries is 0.87–0.96/1000 hours. This epidemiological data show that hamstring injuries are the source of a significant absence for the players from both training and competition (Mendiguchia et al. 2012; Orchard et al. 2013). Hamstring injuries are linked to the interaction of numerous risk factors that can be divided into two categories: non-modifiable and modifiable risk factors (Crosier 2004; Kellis et al. 2012; Kellis 2018; Bisciotti et al. 2020).

In the first category we find:

1. Age.
2. Previous hamstring injuries.
3. Belonging to the Afro-Caribbean ethnic group.

The second category includes:

1. Muscular imbalance between hamstrings and quadriceps femoris.
2. Insufficient hamstring eccentric strength.
3. Insufficient warm-up phase.
4. Reduced sarcomere length.
5. Excessively high and/or poorly dosed workload.
6. Lack of muscle flexibility.
7. Excessive pelvis anteversion and insufficient strength of the lumbo-pelvic muscles.

The hamstring injuries—in accordance with the model proposed by UEFA—are classified into four different categories in relation to their severity and, consequently, to the extent of time loss injury (i.e., the number of training or competition days lost due to the injury) (Hägglund et al. 2003; Waldén et al. 2005; Jacobson and Tegner 2007). The UEFA classification (Ekstrand et al. 2011a, b) is the following:

G. N. Bisciotti et al., *Return-to-Play after Lower Limb Muscle Injury in Football*,
https://doi.org/10.1007/978-3-030-84950-4_18

- All the injuries involving an absence from training sessions or competitions between 1 and 3 days (including the day on which the subject suffered the traumatic event) are classified as "minimal."
- The injuries requiring an absence between 4 and 7 days are classified as "mild."
- The injuries involving activity interruption for 8–28 days are considered "moderate."
- The injuries requiring a time loss that exceeds 28 days are defined "severe."

The hamstring injuries, including total and subtotal ruptures, are generally caused by an extreme and forced hip flexion at the same time as a complete knee extension (Kirkland et al. 2008; Sarimo et al. 2008). In some cases (e.g., in water skiing), hip flexion simultaneous to a sudden knee extension can lead to the proximal avulsion of the long head of the hamstring at the level of its proximal insertion on the ischial tuberosity. Sometimes the semitendinosus muscle is involved too. Similar cases have been described in the literature both in water skiing and in bull riders (Sallay et al. 1996a, b; Chakravarthy et al. 2005), but they remain quite rare in football. In such cases, and in the presence of avulsions with retraction greater than 2–3 cm, the choice is between surgical and conservative treatment. These injuries, in addition to being classified as "major" according to the UEFA, can put at serious risk the athlete's career. The interest for the surgical treatment of these injuries has grown in recent years due to the poor results offered by the conservative treatment as reported in the literature (Chakravarthy et al. 2005; Folsom and Larson 2008; Volpi and Bisciotti n.d.).

18.2 Case Report

LT, 24 years old (height 189 cm and weight 86 kg), European Serie A championship football player with the role of striker, came to our medical observation. In a game situation involving a rapid flexion of the hip at the same time, a sudden thigh extension (movement described with the term "half overturned"), suffered a severe injury to the proximal right femoral biceps at the level of the common insertion with the semitendinosus muscle (Fig. 18.1). The injury was classified as third degree following the "Italian consensus conference on guidelines for conservative treatment of lower limb muscle injuries in athletes" classification (Bisciotti et al. 2018). The athlete was immediately placed in unloading by immobilization in flexion with splint for a period of 10 days, after which he began a rehabilitation period of about 30 days at his club. The rehabilitation plan was essentially based on cryotherapy, instrumental therapy, and cautious mobilization. After this period (once he came to our medical observation), the question arose, given the imminent and pressing competitive commitments, whether to continue with the conservative program or to opt for a surgical resolution. From the literature, the outcome superiority of surgical repair performed in acute (i.e., within the first 4 post-injury weeks) compared to surgery performed in a chronic situation is clear. For this reason, we have opted for the continuation of conservative treatment. In our experience, the hamstring injury is conservatively treated in three phases (Bisciotti et al. 2019). Yet, considering that the athlete came to our medical observation about 40 days after the injury event, we decided that jumping to the second phase was appropriate. However, Table 18.1 reports, for didactic purposes, the composition of the first phase for hamstring injury and the transition criteria to be adopted to pass from the first to the second phase.

18.2.1 Phase II (20 Days)

The transition criteria between the first and the second rehabilitation phase were based on the clinical assessment and the imaging examination (Bisciotti et al. 2018):

Clinical and Functional Criteria:

1. Absence of pain during maximum isometric contraction.
2. No pain during passive stretching.
3. No pain during active stretching tests.
4. Full ROM of the hip joint.

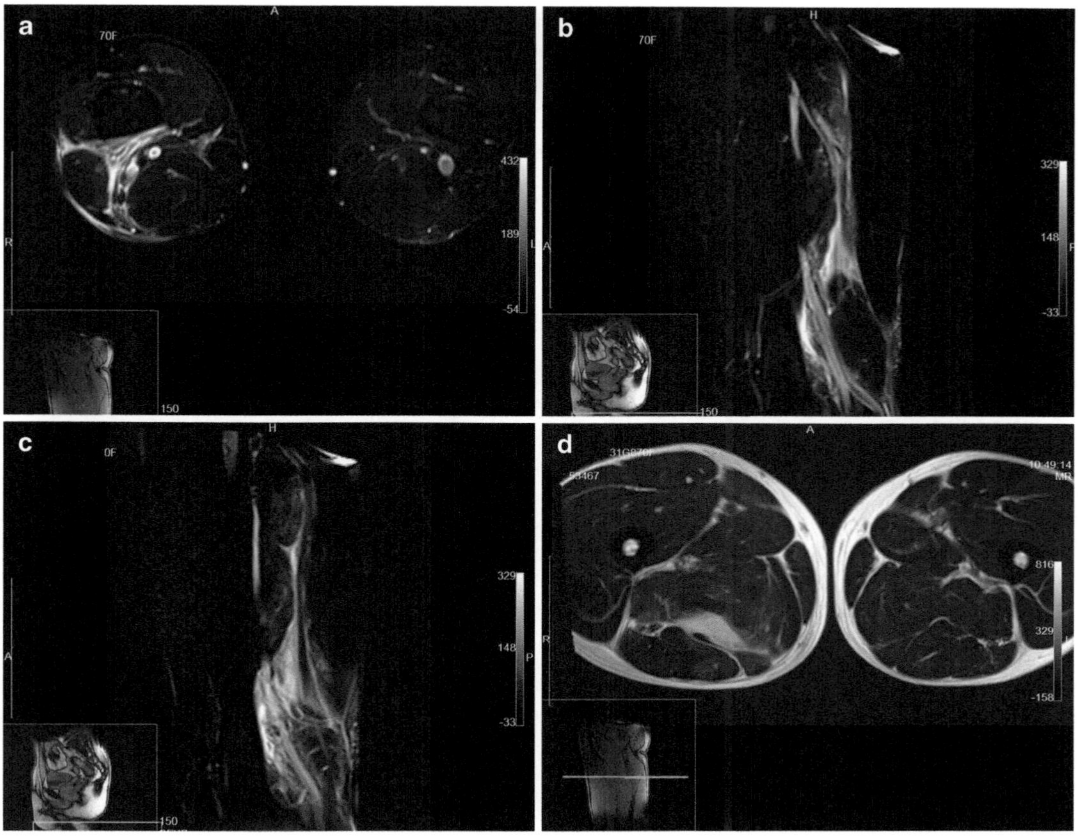

Fig. 18.1 MRI sequence (**a–d**) showing a recent third degree indirect injury (according to the "Italian consensus conference on guidelines for conservative treatment on lower limb muscle injuries in athlete" classification). The injury is located at right biceps femoris proximal insertion in proximity to the common insertion with semitendinosus muscle. It is possible to observe the biceps femoris complete avulsion with a muscle belly retraction of about 2 cm, with peripheral effusion and perimuscular edema. (**a**) axial STIR image. (**b**) sagittal STIR image. (**c**) sagittal STIR image. (**d**) T2 axial image

Imaging Criteria:

1. Decrease of the lesion gap at MRI examination.
2. Presence of granulation repair tissue within the lesion gap in US examination.

Methods Used During the Second Phase.

1. Active and passive stretching.
2. Concentric exercises with progressively increasing intensity.
3. Introduction of straight run with progressive speed increase.
4. Proprioceptive training.

18.2.2 Phase III (30 Days)

The transition criteria between the second and the third rehabilitation phase were the clinical assessment and the imaging examination (Bisciotti et al. 2019):

Clinical and Functional Criteria:

1. No pain during maximal concentric contractions.
2. No pain during submaximal eccentric contractions.

Imaging Criteria:

1. Disappearance of the lesion gap in MRI (or US) examination.

Table 18.1 The first phase for hamstring injury and the transition criteria to be adopted to pass from the first to the second phase

Phase I
Methods used during the first phase:
(1) pain control
(2) mobilization
(3) isometric exercises with progressive intensity increase
(4) Hydrokinesitherapy
The transition criteria between the first and the second rehabilitation phase were the clinical assessment and the imaging examination (Bisciotti et al. 2019):
Clinical and functional criteria:
(1) absence of pain during maximum isometric contraction
(2) no pain during passive stretching
(3) no pain during active stretching tests
(4) full ROM of the hip joint
Imaging criteria:
(1) decrease of the lesion gap at MRI (or US) examination
(2) presence of granulation repair tissue within the lesion gap in US examination

2. Presence of compact granulation repair tissue within the lesion gap in US examination.

Methods Used During the Third Phase.

1. Eccentric exercises with progressively increasing intensity.
2. Run with change of direction movements at progressive speed.
3. Eccentric faster contractions in elongated position.
4. Football individual technique.

The isokinetic exercise progression plan is shown in Table 18.2.

18.3 Clinical Test for Return to Training

The return to training (RTT) decision-making process was based on the guidelines of the Italian consensus conference on return to play after lower limb muscle injury in football (Bisciotti et al. 2019). The following steps in clinical evaluation were respected:

Table 18.2 The isokinetic work program in eccentric modality (performed with an isokinetic device), is introduced during the third phase. The work plan is divided into five phases during which the execution speed is progressively increased until reaching 60 °/s

Phase 1	Speed: 20°/s Series: From 5 to 10 Recovery: 1' ROM and eccentric strength: Gradual increase both of the ROM and eccentric strength *To be maintained for a minimum of three training sessions*
Phase 2	Speed: 30°/s Series: From 5 to 10 Recovery: 1' ROM and eccentric strength: Full ROM achievement. Increase of eccentric strength *To be maintained for a minimum of three training sessions*
Phase 3	Speed: 40°/s Series: From 5 to 10 Recovery: 1' ROM and eccentric strength: Full ROM. Maximum eccentric force *To be maintained for a minimum of three training sessions*
Phase 4	Speed: 50°/s Series: From 5 to 10 Recovery: 1' ROM and eccentric strength: Full ROM. Maximum eccentric force *To be maintained for a minimum of three training sessions*
PHASE 5	Speed: 60°/s Series: From 5 to 10 Recovery: 1' ROM and eccentric strength: Full ROM. Maximum eccentric force *To be maintained for a minimum of three training sessions*

1. Absence of clinical symptoms.
2. Absence of pain or tenderness during muscle palpation.
3. Absence of pain on passive and active stretching.
4. Absence of pain on isometric, concentric and eccentric contraction.
5. Completion of the prescribed rehabilitation program.
6. MRI and US imaging assessment.
7. Subjective feelings of the player were taken into account.

18.3.1 Specific Assessment

The specific assessment for hamstring muscle injury was based on the following tests:

1. Passive straight leg raise test (Witvrouw et al. 2003; Delvaux et al. 2014; Kellis et al. 2015).
2. Dynamic flexibility H test (Askling et al. 2010).

18.4 Laboratory Tests for RTT

In agreement with the Italian consensus statement on return to play after lower limb muscle injury in football (Bisciotti et al. 2019), the evaluation of hamstring muscle strength by dynamometric tests in isometric, isotonic, and isokinetic conditions was performed (Croisier et al. 2002; Sanfilippo et al. 2013; Delvaux et al. 2014; Bisciotti et al. 2016).

18.5 Field Tests for RTT

In agreement with the Italian consensus statement on return to play after lower limb muscle injury in football (Bisciotti et al. 2019), the following tests for RTT decision-making process were performed:

1. Illinois Agility Test (Hachana et al. 2013; Raya et al. 2013; Bisciotti 2015; Negra et al. 2017; Bisciotti et al. 2018).
2. Braking test (Bisciotti 2015; Bisciotti et al. 2018).
3. Backward running (Brumitt et al. 2013).

18.6 Return to Play Tests

Once the player received clearance for RTT, in agreement with the Italian consensus statement on return to play after lower limb muscle injury in football (Bisciotti et al. 2019), the return to play (RTP) decision-making process was started. The process was based on GPS data acquisition. The duration of the process was 20 days during which a certain number of parameters was recorded and subdivided in the following categories:

1. Quantitative evaluation.
2. Qualitative evaluation.
3. Parameter analysis.

For an in-depth description of the parameters and categories, see Chap. 4.

After 20 days, the clearance for RTP was obtained. About 18 months after the injury, no reinjury occurred, and the subject is fully satisfied with the rehabilitation path.

18.7 Discussion

Although hamstring injuries are particularly frequent in sport traumatology, their complete proximal rupture is a rare occurrence. To date, there is no consensus on the treatment of this type of injury, which includes conservative treatment or surgical repair (Chakravarthy et al. 2005). However, in recent years for the complete or near-complete high-grade injuries of the proximal free tendon origin, surgical treatment is generally recommended. According to the majority of authors, surgical repair is indicated only in the case of avulsion trauma with retraction greater than 2–3 cm (Cohen and Bradley 2007; Carmichael et al. 2009; Cohen et al. 2012; Volpi and Bisciotti n.d.). Unfortunately, diagnosis is often delayed both for the rarity of this type of event and for the fact that the depth of the lesion site and the presence of a more or less conspicuous layer of adipose tissue and the belly of the gluteus maximus muscle make the ultrasound exploration quite difficult. Therefore, in the case of suspicion of total proximal rupture, the MRI examination is essential in order to obtain a correct diagnosis (Cohen and Bradley 2007; Sarimo et al. 2008). Acute surgical repair shows a better outcome, in terms of patient satisfaction, strength recovery, and level of return to play, in comparison to delayed surgical repair, that is, the surgical repair performed in chronic condition (Harris et al. 2011). Chronic repair is more technically demanding. Indeed, in a delayed surgical repair, there is an increased need for sciatic nerve neurolysis and larger surgical incisions (Orava and Kujala 1995; Brucker and Imhoff 2005; Cohen and Bradley 2007; Kwak et al. 2011). On the other hand, acute surgical repair reduces both the rate of complication and of reinjury. In literature, the threshold time for "chronic" (or delayed) repair is in general, albeit arbitrarily, fixed between 4 weeks and 3–4 months. It is clear that, because of this very large window of time, the hamstring atrophy may or may not be a component of the clinical situation (Sallay et al. 1996a, b; Slavotinek and Verrall 2002; Rubin 2012). Therefore, our choice to opt for the continuation of conservative treatment was basically dictated by this series of reasons. The third phase of the conservative treatment program adopted was based on isokinetic eccentric conditioning and on the Eccentric Fast Contraction in Elongated position method (EFCEP). Since the rationale for the application of the eccentric method is universally known, we consider appropriate to elucidate the EFCEP method. In the literature there are some studies showing that hamstring injuries, occurring during the sprint, occur during the final part of the swing phase of the gait cycle (Heiderscheit et al. 2005; Schache et al. 2009). Indeed, during the second part of the swing phase, the hamstrings are strongly electrically activated and, at the same time, strongly elongated. Furthermore, in this particular moment of the swing phase, the hamstring muscles are absorbing the energy deriving from the limb deceleration to prepare for the foot contact with the ground (Chumanov et al. 2007; Yu et al. 2008). For these reasons, during the last part of the swing phase, the hamstrings are in rapid transition phase, going from an eccentric contraction (performed in order to decelerate the extension of the knee) to a concentric phase leading them to assume the role of active hip extensors (Hawkins et al. 2001; Drezner 2003), performed in order to decelerate the knee extension. This rapid change, from an eccentric to a concentric phase, makes the hamstring extremely exposed to the injury event (Verrall et al. 2005). In this situation, most of the tension load is absorbed by the biceps femoris long head (Thelen et al. 2005). In this phase, the biceps femoris long head shows greater vulnerability than the semitendinosus and the semimembranosus muscles. The hamstring injuries occurring during the sprint, or in any case in the high-speed running phases, are usually observed at the level of the biceps femoris intramuscular tendon and the adjacent muscle fibers (Koulouris and Connell 2003; Askling et al. 2007). Beyond these general clarifications, however, it is necessary to specifically address the problem concerning the different biomechanical roles played by the biceps femoris (BF) and the semimembranosus muscle (SM). Indeed, even if evaluated globally, the hamstring complex can be considered active from the mid swing up to the terminal stance phase, and the following must be considered:

1. During the mid-swing phase just to the terminal stance phase, the BF is the muscle of the hamstring complex undergoing the greatest elongation. This elongation is on average equal to 12% of its rest length (Schache et al. 2009).
2. Always during the mid-swing phase just to the terminal stance phase, the SM is the muscle of the hamstring complex producing the greatest peak force and absorbing the maximum ratio of the power production (Schache et al. 2009).

Therefore, it is reasonable to think that the biomechanical reasons for BF and SM injuries are substantially different (Askling et al. 2003, 2007). This difference in the injury mechanism etiology may justify a difference in the rehabilitation programs. Therefore, the rehabilitation of the BF should be based on its stretch conditioning, while SM rehabilitation and prevention should focus above all on its strengthening. Askling et al. (2013) demonstrated the best outcome for a BF rehabilitation program based on lengthening, compared to a classic type pro-

Fig. 18.2 The pendulum

Fig. 18.3 The glider

gram. Indeed, the muscle injury may produce a change in the strength-length relationship of the injured muscle, shifting its optimal tension-dynamic behavior toward shorter lengths. The cause of this change could also be attributable to the presence of an excessive amount of post-injury fibrotic tissue (Kaariainen et al. 2000). Therefore, a decrease in serial compliance of the injured muscle (caused by an excessive fibrotic area that is stiffer than the intact muscle tissue) would be able to shorten the length at which the muscle produces its peak force (Brockett et al. 2001). For this reason, the rehabilitation plan should be able to restore an optimal muscle length at which the muscle reaches its force peak, thus reducing the reinjury risk (Witvrouw and Wangesten 2013). Therefore, the EFCEP method is based on eccentric strengthening exercises performed in a situation of muscle lengthening. The exercises must be similar to the performance model and involve both the knee and the hip joint (Chumanov et al. 2007; Schache et al. 2012). This working methodology, based on the principle of "eccentric faster contraction in elongated position," provides that the contractions must necessarily be fast, in order to be as much as possible similar to the injury etiopathogenesis model. Below are some examples of exercises, compliant to the EFCEP principle, that are particularly suitable for the BF rehabilitation (Figs. 18.2, 18.3, 18.4, 18.5, and 18.6).

Fig. 18.4 The diver

Fig. 18.5 The throw forward

Fig. 18.6 Retro walk and retro run

18.8 Conclusions

The complete proximal rupture of the hamstrings is a rare but severe event, which may be career-ending for the athlete. The choice between conservative treatment and surgical intervention must be carefully considered, basing the judgment criteria on different factors such as the athlete's age, their sporting level, and the time elapsed since the injury event. In any case, rehabilitation after a hamstring injury must necessarily consider the different biomechanical roles played by the BF and the SM, and rehabilitation plan should be oriented accordingly.

References

Askling C, Karlsson J, Thorstensson A. Hamstring injury occurrence in elite soccer players after preseason strength training with eccentric overload. Scand J Med Sci. 2003;13:244–50.

Askling CM, Nilsson J, Thorstensson A. A new hamstring test to complement the common clinical examination before return to sport after injury. Knee Surg Sports Traumatol Arthrosc. 2010;18:1798–803.

Askling M, Tengvar M, Thortensson A. Acute hamstring injuries in Swedish elite football. A prospective randomized controlled clinical trial comparing two rehabilitation protocols. Br J Sport Med. 2013 Oct;47(15):953–9.

Askling CM, Tengvar M, Saartok T, Thorstensson A. Acute first-time hamstring strains during high-speed running: a longitudinal study including clinical and magnetic resonance imaging findings. Am J Sports Med. 2007;35:197–206.

Bisciotti GN, Chamari K, Cena E, Carimati G, Bisciotti AL, Bisciotti AN, Quaglia A, Volpi P. Hamstring injuries prevention in soccer: a narrative review of current literature. Joints. 2020;7:215–26.

Bisciotti GN, Quaglia A, Belli A, et al. Return to sports after ACL reconstruction: a new functional test protocol. Muscles Ligaments Tendons J. 2016;06:499–509.

Bisciotti GN, Volpi P, Alberti G, Aprato A, Artina M, Auci, et al. Italian consensus statement (2020) on return to play after lower limb muscle injury in football (soccer). BMJ Open Sport Exerc Med. 2019 Oct 15;5(1):e000505.

Bisciotti GN, Volpi P, Amato M, et al. Italian consensus conference on guidelines for conservative treatment on lower limb muscle injuries in athlete. BMJ Open Sport Exerc Med. 2018;4:e000323. https://doi.org/10.1136/bmjsem-2017-000323.

Bisciotti GN. Return to play after a muscle lesion. In: Volpi P, editor. Arthroscopy in sport. Cham: Springer Edition; 2015.

Brockett CL, Morgan DL, Proske U. Human hamstring muscles adapt to eccentric exercise by changing optimum length. Med Sci Sports Exerc. 2001;33:783–90.

Brucker PU, Imhoff AB. Functional assessment after acute and chronic complete ruptures of the proximal hamstring tendons. Knee Surg Sports Traumatol Arthrosc. 2005;13(5):411–8.

Brumitt J, Heiderscheit BC, Manske RC, et al. Lower extremity functional tests and risk of injury in division III collegiate athletes. Int J Sports Phys Ther. 2013;8:216–27.

Carmichael J, Packham I, Trikha SP, Wood DG. Avulsion of the proximal hamstring origin: surgical technique. J Bone Joint Surg Am. 2009;91(suppl 2):249–56.

Chakravarthy J, Ramisetty N, Pimpalnerkar, Mohtadi N. Surgical repair of complete proximal hamstring tendon ruptures in water skiers and bull riders. A report of four cases and review of the literature. Br J Sport Med. 2005;39:569–72.

Chumanov ES, Heiderscheit BC, Thelen DG. The effect of speed and influence of individual muscles on hamstring mechanics during the swing phase of sprinting. J Biomech. 2007;40:3555–62.

Cohen S, Bradley J. Acute proximal hamstring rupture. J Am Acad Orthop Surg. 2007;15(6):350–5.

Cohen SB, Rangavajjula A, Vyas D, Bradley JP. Functional results and outcomes after repair of proximal hamstring avulsions. Am J Sports Med. 2012;40(9):2092–8.

Croisier J-L, Forthomme B, Namurois M-H, et al. Hamstring muscle strain recurrence and strength performance disorders. Am J Sports Med. 2002;30:199–203.

Crosier J-L. Factors associated with recurrent hamstring injuries. Sports Med. 2004;34:681–95.

Delvaux F, Rochcongar P, Bruyere O, et al. Return-to-play criteria after hamstring injury: actual medicine practice in professional soccer teams. J Sports Sci Med. 2014;13:721–3.

Drezner JA. Practical management: hamstring muscle injuries. Clin J Sport Med. 2003;13:48–52.

Eirale C, Ekstrand J. Epidemiology of injury in football. ASPETAR Sport Med J. 2013;2:144–9.

Ekstrand J, Hägglund M, Walden M. Epidemiology of muscle injuries in professional football (soccer). Am J Sports Med. 2011a;39:1226–32.

Ekstrand J, Hägglund M, Walden M. Injury incidence and injury patterns in professional football: the UEFA injury study. Br J Sports Med. 2011b;45:553–8.

Folsom GJ, Larson CM. Surgical treatment of acute versus chronic complete proximal hamstring ruptures: results of a new allograft technique for chronic reconstructions. Am J Sports Med. 2008;36(1):104–9.

Hachana Y, Chaabene H, Nabli MA, et al. Test-retest reliability, criterion-related validity, and minimal detectable change of the Illinois agility test in male team sport athletes. J Strength Cond Res. 2013;27:2752–9.

Hägglund M, Waldén M, Ekstrand J. Exposure and injury risk in Swedish elite football: a comparison between season 1982 and 2002. Scand J Med Sci Sports. 2003;13:364–70.

Harris JD, Criesser MJ, Best TM, Ellis TJ. Treatment of proximal hamstring ruptures - a systematic review. Int J Sports Med. 2011;32(7):490–5.

Hawkins RD, Hulse MA, Wilkinson C, et al. The association football medical research programme: an audit of injuries in professional football. Br J Sports Med. 2001;35:43–7.

Heiderscheit BC, Hoerth DM, Chumanov ES, Swanson SC, Thelen BJ, Thelen DG. Identifying the time of occurrence of a hamstring strain injury during treadmill running: a case study. Clin Biomech (Bristol, Avon). 2005;20:1072–8.

Jacobson I, Tegner Y. Injuries among Swedish female elite football players: a prospective population study. Scand J Med Sci Sports. 2007 Feb;17(1):84–91.

Kaariainen M, Jarvinen T, Jarvinen M, Rantanen J, Kalimo H. Relation between myofibers and connective tissue during muscle injury repair. Scand J Med Sci Sports. 2000;10:332–7.

Kellis E. Intra- and inter-muscular variations in hamstring architecture and mechanics and their implications for injury: a narrative review. Sports Med. 2018 Oct;48(10):2271–83.

Kellis E, Ellinoudis A, Kofotolis N. Hamstring elongation quantified using ultrasonography during the straight leg raise test in individuals with low back pain. PM R. 2015;7:576–83.

Kellis E, Galanis N, Kapetanos G, Natsis K. Architectural differences between the hamstring muscles. Electromyogr Kinesiol. 2012 Aug;22(4):520–6.

Kirkland A, Garrison JC, Singleton SB, Rodrigo J, Boettner F, Stuckey S. Surgical and therapeutic management of a complete proximal hamstring avulsion after failed conservative approach. J Orthop Sports Phys Ther. 2008;38:754–60.

Koulouris G, Connell D. Evaluation of the hamstring muscle complex following acute injury. Skelet Radiol. 2003;32:582–9.

Kwak HI, Bae SW, Choi YS, Jang MS. Early surgical repair of acute complete rupture of the proximal hamstring tendons. Clin Orthop Surg. 2011;3:249–53.

Mendiguchia J, Alentorn-Geli E, Brughelli M. Hamstring strain injuries: are we heading in the right direction? Br J Sports Med. 2012 Feb;46(2):81–5.

Negra Y, Chaabene H, Hammami M, et al. Agility in young athletes: is it a different ability from speed and power? J Strength Cond Res. 2017;31:727–35.

Orava S, Kujala UM. Rupture of the ischial origin of the hamstring muscles. Am J Sports Med. 1995;23(6):702–5.

Orchard JW, Seward H, Orchard JJ. Results of 2 decades of injury surveillance and public release of data in the Australian football league. Am J Sports Med. 2013 Apr;41(4):734.

Raya MA, Gailey RS, Gaunaurd IA, et al. Comparison of three agility tests with male servicemembers: Edgren side step test, t-test, and Illinois agility test. J Rehabil Res Dev. 2013;50:951–60.

Rubin DA. Imaging diagnosis and prognostication of hamstring injuries. AJR Am J Roentgenol. 2012 Sep;199(3):525–33.

Sallay PI, Friedman RL, Coogan PG, Garrett WE. Hamstring muscle injuries among water skiers. Functional outcome and prevention. Am J Sports Med. 1996a;24:130–6.

Sallay PI, Friedman RL, Coogan PG. Hamstring muscle injury among water skiers: functional outcome and prevention. Am J Sports Med. 1996b;24:137–43.

Sanfilippo JL, Silder A, Sherry MA, et al. Hamstring strength and morphology progression after return to sport from injury. Med Sci Sports Exerc. 2013;45:448–54.

Sarimo J, Lempainen L, Mattila K, Orava S. Complete proximal hamstring avulsions: a series of 41 patients with operative treatment. Am J Sports Med. 2008;36:1110–5.

Schache AG, Dorn TW, Blanch PD, Brown NA, Pandy MG. Mechanics of the human hamstring muscles during sprinting. Med Sci Sports Exerc. 2012 Apr;44(4):647–58.

Schache AG, Wrigley TV, Baker R, Pandy MG. Biomechanical response to hamstring muscle strain injury. Gait Posture. 2009 Feb;29(2): 332–8.

Slavotinek JP, Verrall GM, Fon GT. Hamstring injury in athletes: using MR imaging measurements to compare extent of muscle injury with amount of time lost from competition. AJR Am J Roentgenol. 2002 Dec;179(6):1621–8.

Thelen DG, Chumanov ES, Hoerth DM, et al. Hamstring muscle kinematics during treadmill sprinting. Med Sci Sports Exerc. 2005;37:108–14.

Verrall GM, Slavotinek JP, Barnes PG. The effect of sports specific training on reducing the incidence of hamstring injuries in professional Australian rules football players. Br J Sports Med. 2005;39:363–8.

Volpi P, Bisciotti GN. Muscle injury and regeneration: surgical approach. Min Orth. In press

Waldén M, Hägglund M, Ekstrand J. UEFA champions league study: a prospective study of injuries in professional football during the 2001-2002 season. Br J Sports Med. 2005 Aug;39(8):542–6.

Witvrouw E, Danneels L, Asselman P, et al. Muscle flexibility as a risk factor for developing muscle injuries in male professional soccer players. Am J Sports Med. 2003;31:41–6.

Witvrouw E, Wangesten A. Hamstring rehabilitation. Science or expert practice? ASPETAR Sport Med J. 2013;2:478–85.

Yu B, Queen RM, Abbey AN, Liu Y, Moorman CT, Garrett WE. Hamstring muscle kinematics and activation during overground sprinting. J Biomech. 2008;41:3121–6.

Case Report: Return to Play and Return to Training after Adductor Injury

19

19.1 Introduction

Injuries to the adductor muscles represent a rather common problem in football traumatology. Indeed, they account for about 23% of all indirect muscle injuries and cause an average of 14 days of time loss injury per season (Ekstrand et al. 2011). Adductor injuries have a peak incidence in the age group between 22 and 30 years. The reinjury rate after a first adductor injury (both acute and chronic) is 18% (Ekstrand et al. 2011). A previous history of injury and a reduced strength of the adductor musculature represent a risk factor for reinjury (Engebretsen et al. 2008, 2010). Indeed, footballers who have sustained an injury to the adductor muscles have a double risk of recurrence (Engebretsen et al. 2010). It is interesting to note that professional hockey players showing adductor strength equal to or less than 80% of the abductor strength at the beginning of the season show an injury risk to the adductor muscles 17 times greater than the normal risk (Engebretsen et al. 2010). Amateur football players showing intrinsic weakness of adductor muscles have a four times greater risk of incurring reinjuries (Engebretsen et al. 2010). Finally, concerning the prevention of adductor muscle injuries, according to some studies focused on hockey players, a stretching program performed in the pre-season period has proved to be effective in reducing adductor muscle injuries throughout the season (Tyler et al. 2002).

19.2 Case Report

19.2.1 Anamnesis

B.F., 22 years old, professional footballer, comes to our medical observation complaining of a possible injury to the left adductor complex, which occurred three days earlier, during a change of direction, made during a sprint. The anamnesis was facilitated, following the ICC guidelines, by a dedicated form to be filled in, which can be reviewed in Table 19.1. The data collected in the anamnestic form raised the suspicion of a high-grade muscle injury at the level of the left adductor complex.

The clinical evaluation was divided, as suggested by the "Italian Consensus Conference on guidelines for the conservative treatment of muscle injuries of the lower limb in sports" (ICC) (Bisciotti et al. 2016) into three distinct phases represented by:

1. Anamnesis
2. Inspection
3. Clinical examination

19.2.2 Inspection

As recommended by the ICC, the inspection was focused on the verification of the following points:

G. N. Bisciotti et al., *Return-to-Play after Lower Limb Muscle Injury in Football*,
https://doi.org/10.1007/978-3-030-84950-4_19

Table 19.1 Anamnestic form for data collection

Injury date	November 5, 2018
Injury anatomical location	Left adductor longus
Was US examination carried out?	YES/*NO*
Was MR examination carried out?	YES/NO
Were any other imaging examination performed?	YES/*NO*
If the answer to the previous question is yes, please specify	Type of examination:
Injury classification	Choose between: Delayed onset muscle soreness Fatigue-induced muscle disorder Indirect grade 0 lesion Indirect grade I lesion Indirect grade II lesion (a, b, *c*) Indirect grade III lesion Minor direct muscle injury Moderate direct muscle injury Severe direct muscle injury
What sensation did the player feel at the time of the trauma?	Choose from the following options: Stiffness Dull pain *Acute pain*
Does the player remember precisely the moment of the injury and was it related to a specific movement/situation?	*YES*/NO
Did a blunt episode occur?	YES/*NO*
Did the player hear a "snap?"	YES/*NO*
Did the player have the sensation of a "pinprick" or a "stab?"	*YES*/NO
Was the pain cramp-like or did it develop progressively?	YES/*NO*
Did the pain arise immediately?	*YES*/NO
Is the pain area very localized?	*YES*/NO
Is the player able to indicate the point precisely?	*YES*/NO
Is the pain over the entire muscle?	YES/NO
Did the player fall at the time of the injury?	*YES*/NO
Was the player able to continue playing or running after the injury?	YES/*NO*
Was the muscle stretched sufficiently (stretching) during the warm-up?	*YES*/NO
Was the warm-up done well?	*YES*/NO
Did the player have the feeling of "heavy legs" before the injury?	YES/*NO*
In the days before the injury, had the player trained on surfaces that were unusual for him (stiff, heavy, or frozen ground, etc.)?	YES/*NO*
Had the player recently changed footwear or adopted/changed orthoses?	YES/*NO*
In the days preceding the injury, had the player included new or unusual exercises in his training sessions?	YES/*NO*
Had the player changed the coach in the last month?	YES/*NO*
Had the player changed the physical trainer in the last month?	YES/*NO*
What was the intensity of training in the last 3 days prior to the injury?	Choose a RPE value between 0 and 10: (*7*)
How many official or friendly matches had the player played in the 15 days prior to the injury?	Specify the number: (*2*)
How many recovery days had the player had on average between matches?	Specify the number: (*7*)
Had the player had previous injuries in the same anatomical location?	YES/*NO*. If YES, indicate the number and date of occurrence
Had the player had previous injuries in the same kinetic chain?	YES/*NO*. If YES, indicate the number and date of occurrence

Table 19.1 (continued)

Injury date	November 5, 2018
Had the player had pain in the anatomical location of the injury in the days prior to the injury?	YES/*NO*
If the answer to the previous question is yes, how many days before?	Specify the number of days
Was there a feverish or infectious state?	YES/*NO* If yes, please specify the kind of problem.
Does the player now have pain in the injured area during the activities of daily living?	*YES*/NO

1. Check for swelling (negative result).
2. Check for visible hematoma (negative result).
3. Verification of the presence of a gap or muscle retractions (negative result).
4. Verification of the presence of changes, even if modest in entity, in the profile of the muscle where the injury is suspected, compared to the contralateral muscle (negative result).

19.2.3 Clinical Examination

The clinical examination, always in agreement with the ICC recommendations, was based on the following points:

1. Palpation: revealing a lesion gap of about 0.5 cm at the proximal myotendinous junction of the adductor longus (AL).
2. Verification of joint range of motion: which was reduced and painful in flexion, abduction, and external rotation of the hip (VAS 7/10).
3. Stretching: which showed that the adductor muscle extensibility was reduced and painful in both active and passive modalities (VAS 8/10).
4. Functional maneuvers: consisting in the squeeze test (ST) in the three modalities indicated in Figs. 19.1, 19.2, and 19.3. The three modalities of ST performed showed a VAS of 7/10, 8/10, and 9/10, respectively.

The clinical examination allowed the diagnosis of severe injury of the left AL. This clinical suspicion was confirmed by the magnetic resonance examination (Fig. 19.4).

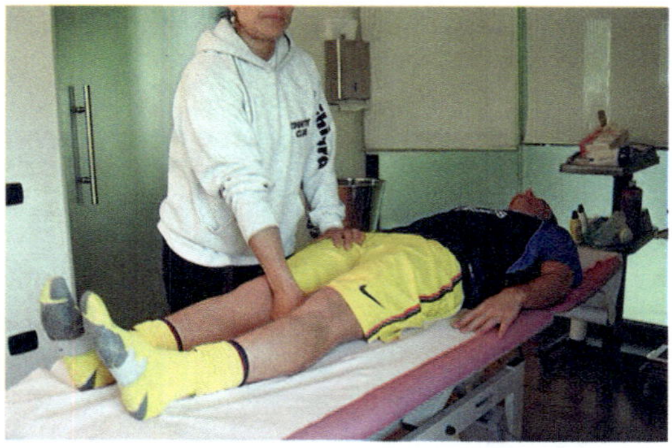

Fig. 19.1 Squeeze test (i.e., isometric adductor contraction) with proximal resistance

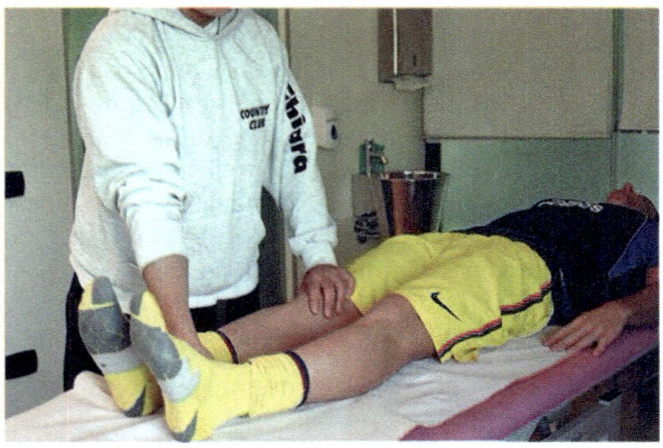

Fig. 19.2 Squeeze test (i.e., isometric adductor contraction) with distal resistance

19.2.4 Therapeutic Path

The conservative therapeutic path was divided into three distinct phases:

Phase I. Based on isometric exercises, lasting 9 days.

The transition criteria between the first and the second rehabilitation phase were based on clini-

Fig. 19.3 Squeeze test (i.e., adductor isometric contraction) with distal resistance and apart legs

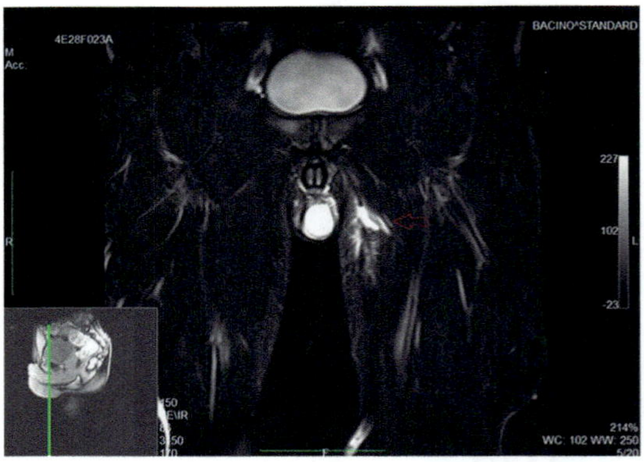

Fig. 19.4 Coronal STIR image showing hyperintense signal alteration at left AL proximal level (arrow). The hyperintensity zone is near the course of the central tendon. The image is in agreement with a second-degree (type c) indirect muscle injury, according to the classification proposed by the "Italian consensus conference on guidelines for conservative treatment on lower limb muscle injuries in athletes" (Bisciotti et al. 2018)

cal assessment and imaging examination (Bisciotti et al. 2018):

Clinical and functional criteria:

1. No pain during maximum isometric contraction
2. No pain during passive stretching
3. No pain during active stretching tests
4. Full ROM of the hip and knee joints

Imaging criteria:

1. Decrease of the lesion gap at US examination.
2. Presence of granulation repair tissue within the lesion gap in US examination.

Phase II. Based on concentric contractions, online running with increasing intensity, active and passive stretching, and proprioceptive training. The phase II lasted 10 days.

The transition criteria between the first and the second rehabilitation phase were based on clinical assessment and imaging examination (Bisciotti et al. 2018):

Clinical and functional criteria:

1. No pain during maximal concentric contractions
2. No pain during submaximal eccentric contractions

Imaging criteria:

1. Disappearance of the lesion gap in MRI examination
2. Presence of compact granulation repair tissue within the lesion gap in US examination

Phase III. Based on:

1. Eccentric exercises of progressive intensity
2. Running with change of direction at progressive intensity
3. Progressive introduction to the specific sport activity

The phase III lasted 25 days.

19.3 Clinical Test for Return to Training

The return to training (RTT) decision-making process was based on ICC guidelines (Bisciotti et al. 2019). The following requirements in clinical evaluation were met:

1. Absence of clinical symptoms.
2. Absence of pain or tenderness during muscle palpation.
3. Absence of pain on passive and active stretching.
4. Absence of pain on isometric, concentric, and eccentric contraction.

5. Completion of the prescribed rehabilitation program.
6. MRI imaging assessment.
7. Subjective feelings of the player were taken into account.

19.3.1 Specific Assessment

The specific assessment for RTT was based on the following clinical tests:

1. Pubic stress test (Hogan and Lovell 1998; Bisciotti 2013; Bisciotti et al. 2016)
2. Resisted hip adduction test (Hogan and Lovell 1998; Croisier et al. 2002; Engebretsen et al. 2010)
3. Squeeze test (Delahunt et al. 2011; Nevin and Delahunt 2014; Bisciotti et al. 2016)
4. Adductor passive stretching test (Atkinson et al. 2010)

19.4 Laboratory Tests for RTT

The laboratory test was based on dynamometric assessment (Croisier et al. 2002; Sanfilippo et al. 2013; Delvaux et al. 2014; Bisciotti et al. 2016).

19.5 Field Tests for RTT

The following field tests were performed:

1. Kicking test (Bisciotti et al. 2016)
2. Carioca test (Kong et al. 2012; Jang et al. 2014)
3. Agility T test (Sassi et al. 2009)

19.6 Return to Play Tests

After receiving the clearance for RTT, the player started the RTP decision-making process based on performance recording via the GPS system. During this period some "typical" sessions were recorded and compared with the same sessions performed during the pre-injury period. The considered parameters were divided in three categories:

1. Quantitative evaluation
2. Qualitative evaluation
3. Parameter analysis

The reference value beyond which the clearance for RTP is postponed was set at a maximum difference of 10% between the pre-injury data and the data recorded during the acquisition period following the RTT.

For an in-depth description of the parameters and categories, see Chap. 4.

After 12 days the clearance for the RTP was obtained.

At 6 months follow-up, no reinjury was recorded and the patient declared to be fully satisfied with the performed rehabilitation process.

19.7 Discussion

The adductor muscle complex includes the adductor longus, the adductor brevis, the adductor magnus, the pectineus, the gracilis, and the external obturator muscle. Among these, the adductor longus (AL) is generally considered to have the greatest influence both in the groin pain syndrome of traumatic origin (Bisciotti et al. 2016; Serner et al. 2017) and in the groin pain syndrome from functional overload (Davis et al. 2012; Bisciotti et al. 2016). From an anatomical point of view, some authors support the hypothesis that more than half of the proximal insertion of the AL is muscular. However, more recently, Davis et al. (2012) found that the AL shows a fibrocartilaginous enthesis and that about 3 mm from its proximal insertion it is still composed for about 90% of tendon tissue. Therefore, the anatomical data remain controversial. In any case, it is well established that the AL has a central tendon. Indeed, its proximal tendon enters into the muscle belly at about 5.5–8 cm from its origin, continuing as an intramuscular tendon, with a total length between 7 and 17 cm (Davis et al. 2012). AL injuries occurring near the proximal, distal, or central tendon have a longer prognosis than purely muscular injuries. This type of injuries requires a longer rehabilitation program (Bisciotti et al. 2015). This case report highlights the strong predictivity of the ST concerning the adductor muscle injuries, which are then confirmed by the MRI examination. The high predictivity positive value (PPV) of the ST (PPV 80–81% [95% CI 63–91]) had already been highlighted by other authors (Serner et al. 2017).

The specific rehabilitation program adopted is justified by the following points:

19.7.1 Scientific Rationale for Phase 1

During the first post-injury days, the lesion gap has not yet developed a sufficiently dense and compact connective tissue scarring. In this period, the greatest risk is that an excessive muscle contraction, especially an eccentric one, may increase the already existing diastasis between the stumps of the damaged fibers. The rehabilitation treatment of the first post-injury days must therefore only include exercises based on submaximal to maximal isometric contractions.

19.7.2 Scientific Rationale for Phase 2

In this phase the granulation tissue gains in compactness and elasticity (Järvinen et al. 2005, 2007, 2014). Therefore, the treatment of this phase must include the exercises based on concentric contraction of progressively increasing intensity.

19.7.3 Scientific Rationale for Phase 3

In this phase, the myofibril interlacing is virtually completed with the interposition of a variable amount of scar tissue (Järvinen et al. 2005, 2007, 2014). Therefore, the introduction of exercises mainly based on eccentric contraction with progressively increasing intensity must be envisaged. Furthermore, it is important to remember that, in the athlete, anatomical healing is not the final stage of the rehabilitation process. Indeed, the anatomical healing process must necessarily be followed by full functional recovery. For this reason, the last rehabilitation phase must necessarily focus on the gradual recovery of the specific performance model.

19.8 Conclusions

This case report underlines the importance of identifying the injury anatomical location. This aspect fundamental for the verification of the lesion proximity to the proximal, distal, or central tendon. Indeed, the prognosis must be based on such anatomical location. Furthermore, the good PPV of the ST concerning the adductor muscle injuries is underlined.

Anamnestic Form

Mr.: B.F.

Age: 22

Sport and role: football (striker)

Assessment date: November 8, 2018

References

Atkinson HDE, Johal P, Falworth MS, et al. Adductor tenotomy: its role in the management of sports-related chronic groin pain. Arch Orthop Trauma Surg. 2010;130:965–70.

Bisciotti GN. La tendinopatia degli adduttoria nel calciatore quando Il ritorno alla corsa? Streng Condition. 2013;5:11–6.

Bisciotti GN, Balzarini L, Volpi P. The classification of muscle injuries; a critical review. Med Sport. 2015;68:165,77.

Bisciotti GN, Volpi P, Zini R, et al. Groin Pain Syndrome Italian Consensus Conference on terminology, clinical evaluation and imaging assessment in groin pain in athlete. BMJ Open Sport Exerc Med. 2016;2:e000142. https://doi.org/10.1136/bmjsem-2016-000142.

Bisciotti GN, Volpi P, Amato M, et al. Italian consensus conference on guidelines for conservative treatment on lower limb muscle injuries in athlete. BMJ Open Sport Exerc Med. 2018;4(1):e000323. https://doi.org/10.1136/bmjsem-2017-000323.

Bisciotti GN, Volpi P, Alberti G, Aprato A, Artina M, Italian consensus statement (2020) on return to play after lower limb muscle injury in football (soccer). BMJ Open Sport Exerc Med. 2019;15;5(1): e000505.

Croisier J-L, Forthomme B, Namurois M-H, et al. Hamstring muscle strain recurrence and strength performance disorders. Am J Sports Med. 2002;30:199–203.

Davis JA, Stringer MD, Woodley SJ. New insights into the proximal tendons of adductor longus, adductor brevis and gracilis. Br J Sports Med. 2012;46(12):871–6.

Delahunt E, McEntee BL, Kennelly C, et al. Intrarater reliability of the adductor squeeze test in gaelic games athletes. J Athl Train. 2011;46:241–5.

Delvaux F, Rochcongar P, Bruyere O, et al. Return-to-play criteria after hamstring injury: actual medicine practice in professional soccer teams. J Sports Sci Med. 2014;13:721–3.

Ekstrand J, Hagglund M, Walden M. Epidemiology of muscle injuries in professional football (soccer). Am J Sports Med. 2011;39:12261232.

Engebretsen AH, Myklebust G, Holme I, Engebretsen L, Bahr R. Prevention of injuries among male soccer players: a prospective, randomized intervention study targeting players with previous injuries or reduced function. Am J Sports Med. 2008;36:1052–60.

Engebretsen A, Myklebust G, Holme I, Engebretsen L, Bahr R. Intrinsic risk factors for groin injuries among male soccer players: a prospective cohort study. Am J Sports Med. 2010;38:2051–7.

Hogan A, Lovell G. The groin pain provocation test. In: Brown A, editor. 4th world football symposium conference proceedings. London: Routledge; 1998.

Jang SH, Kim JG, Ha JK, et al. Functional performance tests as indicators of returning to sports after anterior cruciate ligament reconstruction. Knee. 2014;21:95–101.

Järvinen TA, Järvinen TL, Kääriäinen M, Kalimo H, Järvinen M. Muscle injuries: biology and treatment. Am J Sports Med. 2005;33(5):745–64.

Järvinen TA, Järvinen TL, Kääriäinen M, Äärimaa V, Vaittinen S, Kalimo H, Järvinen M. Muscle injuries: optimising recovery. Best Pract Res Clin Rheumatol. 2007;21(2):317–31.

Järvinen TA, Järvinen M, Kalimo H. Regeneration of injured skeletal muscle after the injury. Muscles Ligaments Tendons J. 2014;3(4):337–45.

Kong DH, Yang SJ, Ha JK, et al. Validation of functional performance tests after anterior cruciate ligament reconstruction. Knee Surg Relat Res. 2012;24:40–5.

Nevin F, Delahunt E. Adductor squeeze test values and hip joint range of motion in Gaelic football athletes with longstanding groin pain. J Sci Med Sport. 2014;17:155–9.

Sanfilippo JL, Silder A, Sherry MA, et al. Hamstring strength and morphology progression after return to sport from injury. Med Sci Sports Exerc. 2013;45:448–54.

Sassi RH, Dardouri W, Yahmed MH, Gmada N, Mahfoudhi ME, Gharbi Z. Relative and absolute reliability of a modified agility T-test and its relationship with vertical jump and straight sprint. J Strength Cond Res. 2009;23(6):1644–51.

Serner A, Roemer FW, Hölmich P, Thorborg K, Niu J, Weir A, Tol JL, Guermazi A. Reliability of MRI assessment of acute musculotendinous groin injuries in athletes. Eur Radiol. 2017;27(4):1486–95.

Tyler TF, Nicholas SJ, Campbell RJ, Donellan S, McHugh MP. The effectiveness of a preseason exercise program to prevent adductor muscle strains in professional ice hockey players. Am J Sports Med. 2002;30:680–3.

20 Case Report: Return to Play and Return to Training After Soleus-Gastrocnemius Injury

20.1 Introduction

The soleus-gastrocnemius muscle complex plays a fundamental role in many sports movements; therefore its injuries have a major impact on many sports activities (Luke et al. 2002; Colberg et al. 2015; Rebella 2015). Soleus-gastrocnemius muscle complex injuries are common in sport activities involving high-speed running, changes of direction, acceleration and deceleration, and jumps. The gastrocnemius plays an important role during the sprint and is an important "power producer" during the movements. Furthermore, the gastrocnemius muscles show a statistically significant correlation with maximal sprinting speed (Abe et al. 2000, 2001). Among the non-modifiable risk factors, the athletes' chronological age and previous muscle injury at soleus-gastrocnemius muscle complex are the stronger risk factors for future muscle injuries (Green and Pizzari 2017). Classically, this type of injuries is known under the common name of "tennis leg" (especially the medial gastrocnemius injuries) because they frequently occur when a tennis player sprints for a ball. The gastrocnemius injuries are usually associated with ballistic movements while the ankle is flexed and the knee is extended (i.e., during an eccentric calf contraction). Usually, the athlete feels a sudden tear or pop with acute pain. The injuries at medial gastrocnemius muscle, and in particular at musculotendinous junction, are the most frequent (Fields and Rigby 2016). Concerning the soleus muscle, about 95% of its fibers are type I (Soukupa et al. 2002), and the muscle's function is to maintain posture and other low-energy activities, such as walking. Furthermore, it is a monoarticular muscle. Despite these two protective factors, the soleus muscle is still exposed to indirect trauma (Balius et al. 2014). Indeed, the soleus shows a medial and lateral intramuscular aponeuroses (Courthaliac and Weilbacher 2007) starting at the anterior epimysium and running distally to the interior of the muscle body (Testut and Latarget 1979; Balius et al. 2013). Furthermore, in the central portion of the muscle, there is an intramuscular tendon that reaches the Achilles tendon. This tendon structure contributes to the Achilles tendon formation. This complicated "myo-connective system complex" explains the muscle's multipennate structure (Chow et al. 2000; Hodgson et al. 2006; Joshi et al. 2010) predisposing it to the injury at the level of any of its musculotendinous junctions (Garrett Jr 1996). Balius et al. (2013) showed five anatomical sites in the soleus muscle where indirect injuries are located, that is:

1. Three musculotendinous junction sites: proximal medial injuries accounting for 25.5% of all injuries, proximal lateral injuries accounting for 12.7%, and distal central tendon injuries accounting for 18.2%.
2. Two myofascial sites: anterior strains 21.8% and posterior strains 21.8%.

G. N. Bisciotti et al., *Return-to-Play after Lower Limb Muscle Injury in Football*,
https://doi.org/10.1007/978-3-030-84950-4_20

Furthermore, in the same study, the authors noted that the proximal medial musculotendinous junction injuries were the most common type of soleus muscle injury (56.4% of all cases).

The most frequent injury risk situation for the soleus muscle is when the knee is flexed. On the contrary, the most frequent injury risk situation for the gastrocnemius muscle is when the knee is in a more extended position.

20.2 Case Report

DH, 25 years old (height 181 cm and weight 70 kg), European Serie A championship football player with the role of defender, came to our medical observation complaining of a possible injury to the left medial gastrocnemius, which occurred 1 day earlier, during a sprint performed in a training session.

The clinical evaluation was divided, as suggested by the "Italian Consensus Conference on guidelines for the conservative treatment of muscle injuries of the lower limb in sports" (ICC) (Bisciotti et al. 2016) into:

1. Anamnesis
2. Inspection
3. Clinical examination

20.2.1 Anamnesis

The player refers a sharp and sudden pain at level of the medial gastrocnemius during a sprint performed in a session training about 24 h before. The player was forced to stop training. Until the next day he had pain putting his body weight on the injured leg and walked with a limp. The pain area was well localized and reproducible.

20.2.2 Inspection

The inspection examination verified the following points:

1. Check for swelling (negative result)
2. Check for visible hematoma (negative result)
3. Check gaps or muscle retractions (negative result)
4. Check for changes in the profile of the injured muscle (negative result)

20.2.3 Clinical Examination

The clinical examination, in agreement with the ICC recommendations, was based on the following points:

1. Palpation: showing a lesion gap of about middle third of the left medial gastrocnemius.
2. Verification of joint range of motion: the range of motion of foot dorsiflexion was reduced and causing pain in the injured area (VAS 8/10).
3. Stretching: the gastrocnemius-soleus muscle complex extensibility was reduced and painful in both active and passive modalities (VAS 8/10).
4. Functional maneuvers: isometric, concentric, and eccentric contractions of gastrocnemius muscle (i.e., plantar dorsiflexion with extended knee joint) caused pain (the VAS was 5/10, – 8/10, – 9/10, respectively).

The clinical examination allowed the diagnosis of injury of the left medial gastrocnemius at middle third level. This clinical suspicion was confirmed by the magnetic resonance examination (Fig. 20.1a, b). The injury, according to ICC classification (Bisciotti et al. 2018), was a first-degree indirect muscle injury.

20.2.4 Therapeutic Path

The conservative therapeutic path was divided into three distinct phases:

Phase I. Based on isometric exercises, lasting 6 days.

The transition criteria between the first and the second rehabilitation phase were based on clinical assessment and imaging examination (Bisciotti et al. 2018).

Clinical and functional criteria:

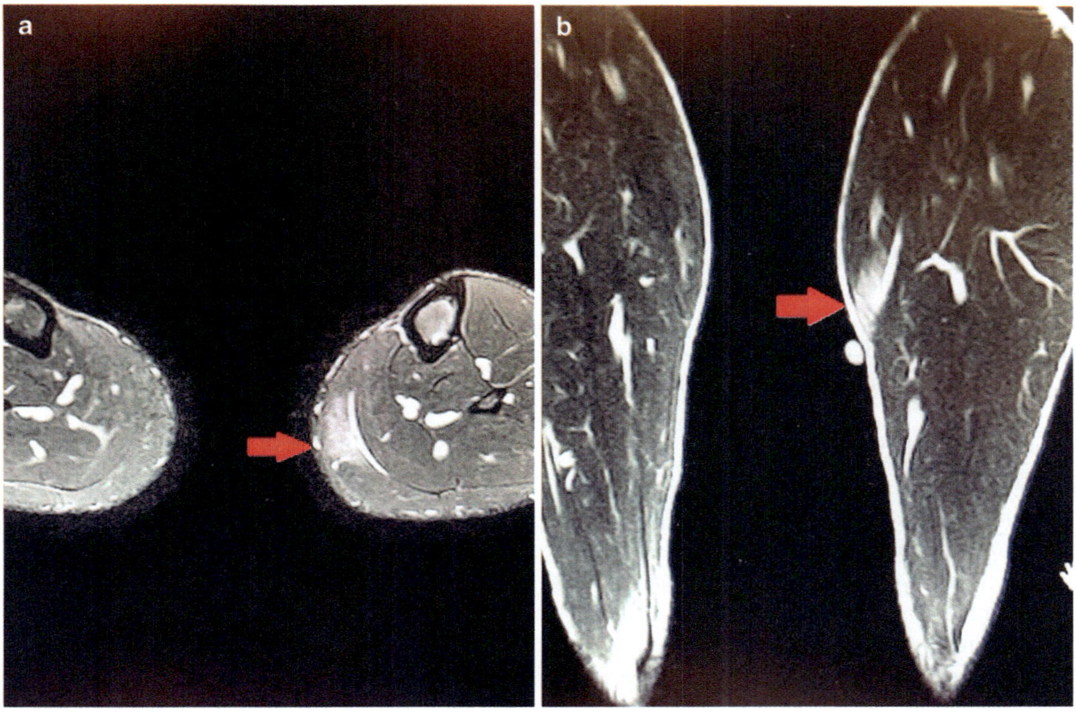

Fig. 20.1 Axial T2 MRI imaging (**a**) and coronal STIR imaging (**b**) showing an injury in the left medial gastrocnemius at its middle third level (arrows). The injury, according to classification proposed by the ICC (Bisciotti et al. 2018), was classified as first-degree indirect muscle injury

1. No pain during maximum isometric contraction
2. No pain during passive stretching
3. No pain during active stretching tests
4. Full ROM of the ankle joint.

Imaging criteria:

1. Decrease of the lesion gap at MRI (or US) examination
2. Presence of granulation repair tissue within the lesion gap in US examination

Phase II. Based on concentric contractions, online running with increasing intensity, active and passive stretching, and proprioceptive training. Phase II lasted 5 days.

The transition criteria between the first and the second rehabilitation phase were based on clinical assessment and imaging examination (Bisciotti et al. 2018):

Clinical and functional criteria:

1. No pain during maximal concentric contractions
2. No pain during submaximal eccentric contractions

Imaging criteria:

1. Disappearance of the lesion gap in MRI examination
2. Presence of compact granulation repair tissue within the lesion gap in US examination

Phase III. Based on:

1. Eccentric exercises of progressive intensity
2. Running with change of direction at progressive intensity
3. Plyometric training.
4. Progressive introduction to the specific sport activity

Phase III lasted 8 days.

20.3 Clinical Test for Return to Training

The return to training (RTT) decision-making process was based on the guidelines of the Italian consensus statement on return to play after lower limb muscle injury in football (Bisciotti et al. 2019). The following requirements in clinical evaluation were met:

1. Absence of clinical symptoms.
2. Absence of pain or tenderness during muscle palpation.
3. Absence of pain on passive and active stretching.
4. Absence of pain on isometric, concentric and eccentric contraction.
5. Completion of the prescribed rehabilitation program.
6. MRI imaging assessment.
7. Subjective feelings of the player were taken into account.

20.3.1 Specific Assessment

The specific assessment for RTT was based on the following clinical tests:

1. Heel-raise test (Moller et al. 2005; Harris-Love et al. 2014)
2. Weight bearing lunge test (Hoch and McKeon 2011; Powden et al. 2015; Baumach et al. 2016)

20.4 Laboratory Tests for RTT

After the validation of the clinical tests, the athlete performed the laboratory test battery.

The laboratory tests performed were:

1. Gastrocnemius muscle strength assessed by dynamometric tests (Croisier et al. 2002; Malliaropoulos et al. 2011; Askling et al. 2010; Sanfilippo et al. 2013)
2. Synchro plates test (Bisciotti et al. 2016)
3. Drop jump test (Silbernagel et al. 2006; Hewett et al. 2007; Powell et al. 2018)

20.5 Field Tests for RTT

After the validation of the laboratory tests, the athlete performed the field test battery.

The field tests performed were:

1. Illinois Agility Test (Hachana et al. 2013; Raya et al. 2013; Bisciotti 2015; Negra et al. 2017; Bisciotti and Volpi 2018).
2. Agility *T*-test (Sassi et al. 2009)

20.6 Return to Play Tests

After the successful conclusion of the field test battery, the athlete moved on to the return to play (RTP) phase. The criteria for RTP decision-making process were based on the GPS data registration.

The data acquisition period started on the first day of RTP and lasted 7 days.

Some "typical" sessions were identified and recorded. The recorded data were compared with the data from similar training sessions of the pre-injury period.

As already mentioned in Chap. 4, the three evaluation categories were:

1. Quantitative evaluation
2. Qualitative evaluation
3. Parameter analysis

The reference values beyond which the clearance for RTP is postponed was set at a maximum difference of 10% between the pre-lesional data and the data recorded during the acquisition period. After 7 days the athlete received clearance for RTP.

20.7 Discussion

Calf injuries are mostly found at gastrocnemius medial head level (Brukner and Khan 2002). This specific type of injury was for the first time described in 1883 in association with tennis. For this reason, this injury is commonly called "tennis leg." Indeed, its paradigmatic presentation is a middle-aged, male amateur tennis player, which injures his medial gastrocnemius suddenly extending the knee with the foot in dorsiflexion position. This typical situation results in immediate pain, disability, and sometimes, depending on the severity, even swelling. The gastrocnemius has a high risk of injury because of its biarticular nature (it crosses both the knee and ankle joints) and its composition including a high percentage of fast-twitch muscle fibers, both of FTa and FTb type. (Garrett Jr 1996; DeLee et al. 2003; Armfield et al. 2006; Simon et al. 2006). This situation of coexisting biarticularity and important percentage of fast-twitch fibers results in a high injury risk. The gastrocnemius injury mechanism evokes the image of a whip snapping in the air. For this reason, the gastrocnemius indirect injuries have historically been called "coup de fouet" or "snap of the whip." The clinical history can help to distinguish between soleus and gastrocnemius injury. However, the clinical examination, based on inspection, palpation, checking of the range of motion, and functional maneuver tests, provides more information (Bisciotti et al. 2018).

During the inspection, the following key points are to be considered:

1. Check for swelling
2. Check for the presence of hematoma (strongly suggestive of the indirect lesion) or ecchymosis (strongly suggestive of direct trauma)
3. Check for the presence of gap or clear muscle retraction
4. Check for changes in the muscular profile, even though modest, in the region of the suspected lesion in comparison to the contralateral muscle

The palpation exam requires the clinician to have specific skills and experience. The palpation exam shall be conducted in two ways:

1. Palpation exam mode 1: the patient is positioned in a way that the examined muscle is in a slightly elongated position (hence slightly contracted eccentrically).
2. Palpation exam mode 2: the muscle to be examined must be completely relaxed.

The palpation exam should be repeated several times, with the two modalities, both on the entire muscle belly and on the area/point of pain indicated by the patient. The clinician should get information, which is obviously subjective, from the skin, subcutaneous tissue, fascia, and muscle. The palpation exam should be performed, when applicable, in a comparative manner and in both modalities described above, exerting a moderate pressure, in a proximal-distal direction and vice versa, always following the orientation of the fibers. At the same time, attention must be paid to the verbal and mimic responses of the patient. At their choice, the operator can perform the palpation exam according to a third mode, in which the patient is asked to perform a slight isotonic contraction of the examined muscle.

The palpation exam aims at verifying:

1. The tone of the muscles affected by the alleged lesion compared to the contralateral muscles. In addition, it is of utmost importance to check the tone of the muscles adjacent to the presumably injured one, as a tone alteration may suggest a high-grade lesion.
2. The presence of gap or clear muscle retraction.
3. The existence of stiff zones.
4. The existence of previous fibrotic areas or altered myo-fascial adhesion.
5. The existence of painful areas or otherwise impaired perception by the patient.
6. The range of motion, which provides the evaluation of the proximal and, when applicable, distal joints to the injured muscle.

Gastrocnemius indirect injuries typically present with tenderness in the medial belly or the musculotendinous junction, while in the soleus indirect injuries the pain is often lateral.

The stretching of the injured muscle should be performed in the following modalities:

1. Active modality
2. Passive modality

Generally, a structural lesion results in pain on both passive and active stretching, while in a functional pathology (such as the DOMS), stretching may provide a pleasant feeling to the patient.

Regarding the functional maneuvers, the injured muscle should be tested in three ways:

1. Maximum isometric contraction
2. Concentric contraction of medium intensity against the operator's resistance
3. Medium effort eccentric contraction against the operator's resistance with no pain

Quantification (from zero to ten) of the perceived pain in accordance with the Visual Analog Scale will be required for each type of contraction. In VAS, the zero value means absence of pain, a value between one and three means a slight pain, a value between four and six means a moderate pain, and a value starting from seven means a severe pain. A value greater than three is suggestive of structural lesion.

From the anatomical point of view, gastrocnemius and soleus have different origins. The medial and lateral head of the gastrocnemius originate from the upper part of the corresponding femoral condyle and from the adjacent part of the knee joint capsule, while the soleus arises from the back of the fibula head and from the soleal line. Such anatomical difference allows the clinician to isolate the activation of the two muscles by varying the degree of knee joint flexion. When keeping the knee flexed at 90°, the soleus muscle is the most important force generator during the plantar flexion, while with the knee in full extension, the gastrocnemius becomes the grater contributor for the plantar flexion (Bojsen-Moller et al. 2004). This functional difference allows the clinician to more carefully test its strength. The same approach should be used during the stretching test. In this case, the knee joint should be placed both in flexed (90°) and full extended position, and the ankle should be dorsiflexed to test the extensibility of soleus and gastrocnemius muscles, respectively. However, it is important to note that a concomitant indirect injury of both soleus and gastrocnemius muscles is possible. Armfield et al. (2006) showed that coexisting injuries in gastrocnemius and soleus muscles were found in 17% of the calf injuries. Concerning the imaging, MRI and musculoskeletal ultrasound (US) are the two choices for investigation (Nquyen et al. 2000). However, US is not a sensitive technique for detecting and assessing soleus injuries compared with MRI (Balius et al. 2014).

20.8 Conclusion

Injuries of the gastrocnemius-soleus muscle complex are relatively common in football. It is important to remember that the gastrocnemius and soleus injuries require different clinical evaluation tests. To successfully complete the rehabilitation process, all the stages of the RTT and the RTP decision-making process must be strictly observed.

References

Abe T, Fukashiro S, Harada Y, et al. Relationship between sprint performance and muscle fascicle length in female sprinters. J Physiol Anthropol Appl Hum Sci. 2001;20:141–7.

Abe T, Kumagai K, Brechue WF. Fascicle length of leg muscles is greater in sprinters than distance runners. Med Sci Sports Exerc. 2000;32:1125–9.

Armfield DR, Kim DH, Towers JD, Bradley JP, Robertson DD. Sports-related muscle injury in the lower extremity. Clin Sports Med. 2006;25:803–42.

Askling CM, Nilsson J, Thorstensson A. A new hamstring test to complement the common clinical examination before return to sport after injury. Knee Surg Sports Traumatol Arthrosc. 2010;18:1798–803.

Balius R, Alomar X, Rodas G, Miguel-Pérez M, Pedret C, Dobado MC, et al. The soleus muscle: MRI, anatomic and histologic findings in cadavers with clinical correlation of strain injury distribution. Skelet Radiol. 2013;42:521–30.

Balius R, Rodas G, Pedret C, Capdevila L, Alomar X, et al. Soleus muscle injury: sensitivity of ultrasound patterns. Skelet Radiol. 2014;43(6):805–12.

Baumach SF, Braunstein M, Regauer M, et al. Diagnosis of musculus gastrocnemius tightness - key factors for the clinical examination. J Vis Exp. 2016;113:53446.

Bisciotti GN, Quaglia A, Belli A, et al. Return to sports after ACL reconstruction: a new functional test protocol. Muscles Ligaments Tendons J 2016;06:499–509.

Bisciotti GN, Volpi P, Alberti G, Aprato A, Artina M,. Italian consensus statement (2020) on return to play after lower limb muscle injury in football (soccer). BMJ Open Sport Exerc Med. 2019;5(1):e000505.

Bisciotti GN, Volpi P, Amato M, et al. Italian consensus conference on guidelines for conservative treatment on lower limb muscle injuries in athlete. BMJ Open Sport Exerc Med. 2018;4(1):e000323. https://doi.org/10.1136/bmjsem-2017-000323.

Bisciotti GN, Volpi P. Return to play. In: Volpi P, editor. Football doctor manual. Trento: Edra Edition; 2018. p. 247–59.

Bisciotti GN. Return to play after a muscle lesion. In: Volpi P, editor. Arthroscopy in sport. Springer; 2015.

Bojsen-Moller J, Hansen P, Aagaard P, Svantesson U, Kjaer M, et al. Differential displacement of the human soleus and medial gastrocnemius aponeuroses during isometric plantar flexor contractions in vivo. J Appl Physiol. 2004;97:1908–14.

Brukner P, Khan K. Clinical sports medicine. 2nd ed. North Ryde: McGraw-Hill; 2002.

Chow RS, Medri MK, Martin DC, Leekam RN, Agur AM, McKee NH. Sonographic studies of human soleus and gastrocnemius muscle architecture: gender variability. Eur J Appl Physiol. 2000;82:236–44.

Colberg RE, Aune KT, Choi AJ, et al. Incidence and prevalence of musculoskeletal conditions in collegiate tennis athletes. JMST. 2015;20:137–44.

Courthaliac C, Weilbacher H. Imaginerie du mollet douloureux chez le sportif. J Radiol. 2007;88:200–8.

Croisier J-L, Forthomme B, Namurois M-H, et al. Hamstring muscle strain recurrence and strength performance disorders. Am J Sports Med. 2002;30:199–203.

DeLee JC, Drez D Jr, Miller MD, editors. DeLee & Drez's orthopaedic sports medicine; principles and practice. 2nd ed. Philadelphia: Saunders; 2003.

Fields KB, Rigby MD. Muscular calf injuries in runners. Curr Sports Med Rep. 2016;15(5):320–4.

Garrett WE Jr. Muscle strain injuries. Am J Sports Med. 1996;24:S2–8.

Green B, Pizzari T. Calf muscle strain injuries in sport: a systematic review of risk factors for injury. Br J Sports Med. 2017;51(16):1189–94.

Hachana Y, Chaabene H, Nabli MA, et al. Test-retest reliability, criterion-related validity, and minimal detectable change of the Illinois agility test in male team sport athletes. J Strength Cond Res. 2013;27:2752–9.

Harris-Love MO, Shrader JA, Davenport TE, et al. Are repeated single-limb heel raises and manual muscle testing associated with peak plantar-flexor force in people with inclusion body myositis? Phys Ther. 2014;94:543–52.

Hewett T, Snyder-Mackler L, Spindler KP. The drop-jump screening test: difference in lower limb control by gender and effect of neuromuscular training in female athletes. Am J Sports Med 2007;35:145.

Hoch MC, McKeon PO. Normative range of weight-bearing lunge test performance asymmetry in healthy adults. Man Ther. 2011;16(5):516–9.

Hodgson J, Finni T, Lai A, et al. Influence of structure on the tissue dynamics of the human soleus muscle observed in MRI studies during isometric contractions. J Morphol. 2006;267:584–601.

Joshi SS, Joshi SD, Kishve PS. Feathered fibres of human soleus. Int J Morphol. 2010;28:239–42.

Luke AC, Kinney SA, D'Hemecourt PA, et al. Determinants of injuries in young dancers. Med Probl Perform Art. 2002;17:105–12.

Malliaropoulos N, Isinkaye T, Tsitas K, et al. Reinjury after acute posterior thigh muscle injuries in elite track and field athletes. Am J Sports Med. 2011;39:304–10.

Moller M, Lind K, Styf J, et al. The reliability of isokinetic testing of the ankle joint and a heel-raise test for endurance. Knee Surg Sports Traumatol Arthrosc. 2005;13:60–71.

Negra Y, Chaabene H, Hammami M, et al. Agility in young athletes: is it a different ability from speed and power? J Strength Cond Res. 2017;31:727–35.

Nquyen B, Brandser E, Rubin DA. Pains, strains and fasciculations: lower extremity muscle disorders. Magn Reson Imaging Clin N Am. 2000;8(2):391–408.

Powden CJ, Hoch JM, Hoch MC. Reliarbility and minimal detectable change of the weight-bearing lunge test: a systematic review. Man Ther. 2015;20(4):524–32.

Powell HC, Silbernagel KG, Brorsson A, et al. Individuals post Achilles tendon rupture exhibit asymmetrical knee and ankle kinetics and loading rates during a drop countermovement jump. J Orthop Sports Phys Ther. 2018;48:34–43.

Raya MA, Gailey RS, Gaunaurd IA, et al. Comparison of three agility tests with male servicemembers: Edgren side step test, t-test, and Illinois Agility test. J Rehabil Res Dev. 2013;50:951–60.

Rebella G. A prospective study of injury patterns in collegiate Pole Vaulters. Am J Sports Med. 2015;43:808–15.

Sanfilippo JL, Silder A, Sherry MA, et al. Hamstring strength and morphology progression after return to sport from injury. Med Sci Sports Exerc. 2013;45:448–54.

Sassi RH, Dardouri W, Yahmed MH, Gmada N, Mahfoudhi ME, Gharbi Z. Relative and absolute reliability of a modified agility T-test and its relationship with vertical jump and straight sprint. J Strength Cond Res. 2009;23(6):1644–51.

Silbernagel KG, Gustavsson A, Thomeé R, et al. Evaluation of lower leg function in patients with Achilles tendinopathy. Knee Surg Sports Traumatol Arthrosc. 2006;14:1207–17.

Simon RR, Sherman SC, Koenigsknecht SJ, editors. Emergency orthopedics: the extremities. 5th ed. New York: McGraw-Hill; 2006.

Soukupa T, Zachařováa G, Smerdub V. Fibre type composition of soleus and extensor digitorum longus muscles in normal female inbred Lewis rats. Acta Histochem. 2002;104(4):399–405.

Testut L, Latarget A. Anatomía Humana. 9th ed. Barcelona: Salvat Editores; 1979. p. 1149–98.

Case Report: Return to Play and Return to Training After Hip Short External Rotator Muscles Injury

21

21.1 Introduction

This chapter describes a case of acute groin pain syndrome (GPS) due to internal obturator muscle injury in a professional football player. GPS, as proposed by the "Groin Pain Syndrome Italian Consensus Conference on Terminology, Clinical Evaluation and Imaging Assessment in Groin Pain in Athlete" (Bisciotti et al. 2016), can be defined as "Any clinical symptom reported by the patient, located at the inguinal-pubic-adductor area, affecting sports activities and/or interfering with the activities of daily living and requiring medical attention."

GPS can be divided into three main categories:

1. GPS due to functional overload, which is characterized by insidious and progressive onset, without an acute trauma, or any other situation which the onset of pain symptoms can be attributed to with certainty.
2. GPS of traumatic origin, in which the onset of pain corresponds to any acute trauma, and such hypothesis is supported by medical history, clinical examination, and imaging.
3. Long-standing GPS or chronic GPS, in which the cohort of symptoms reported by the patient continues for a long period (over 12 weeks) and is recalcitrant to any conservative therapy.

Traumatic GPS is usually due to acute muscle-tendon problems, and the most affected muscle groups are rectus abdominis, adductors, and iliopsoas muscle (Bisciotti et al. 2016). The internal obturator muscle (IO) lesion is very rare, and to our knowledge, there are only six studies concerning IO injuries in literature (Khodaee et al. 2015; Kelm et al. 2016; Wong-On et al. 2018). In this chapter, a case of a traumatic GPS caused by an indirect injury at the level of IO in a professional football player is presented.

21.2 Case Report

A 29-year-old professional male soccer player came to our clinical evaluation complaining GPS at the anterior level of the left hip that radiated obliquely and medially downward. The patient referred the onset of pain after a violent hip intra-rotation movement, with hip and knee flexed at about 90°, during a training session 5 days earlier. Passive internal rotation maneuvers with and without muscle opposition at 90° knee flexion caused pain (VAS 6/10). Examination of MRI (Fig. 21.1) showed a I° grade lesion (LMIn I°), according to the classification proposed by the Italian Consensus Conference on guidelines for conservative treatment on lower limb muscle injuries in athletes (Bisciotti et al. 2019), at the left internal obturator level. The patient was

G. N. Bisciotti et al., *Return-to-Play after Lower Limb Muscle Injury in Football*,
https://doi.org/10.1007/978-3-030-84950-4_21

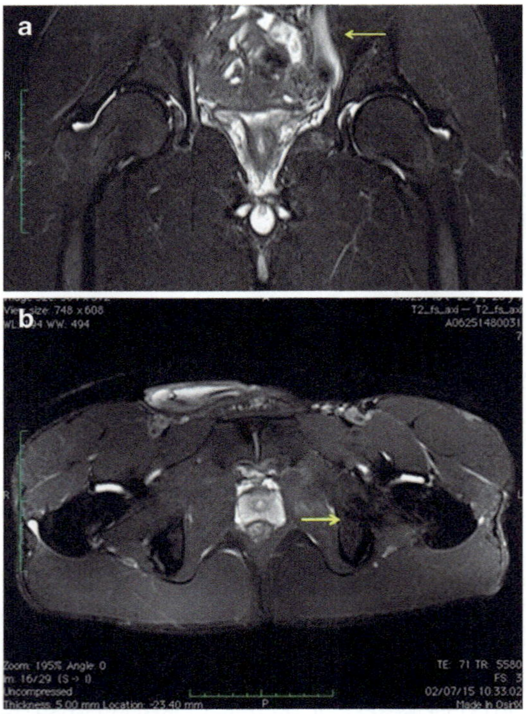

Fig. 21.1 Coronal STIR MRI imaging (**a**) and axial T2 MRI imaging (**b**) showing an indirect muscle first-degree injury (LMIn I °) at the left IO (arrows)

assigned to a rehabilitation path composed of three phases (Bisciotti et al. 2019).

21.2.1 Phase I (5 days)

Methods used during the first phase

1. Pain control
2. Mobilization
3. Isometric exercises with progressive intensity increase
4. Hydrokinesitherapy

21.2.2 Phase II (7 Days)

The transition criteria from the first to the second rehabilitation phase were clinical assessment and imaging examination (Bisciotti et al. 2019):

Clinical and functional criteria:

1. Absence of pain during maximum isometric contraction.
2. No pain during passive stretching.
3. No pain during active stretching tests.
4. Full ROM of the hip joint.

Imaging criteria:

1. Decrease of the lesion gap at MRI (or US) examination.
2. Presence of granulation repair tissue within the lesion gap in US examination.

Methods used during the second phase

1. Active and passive stretching
2. Concentric exercises with progressive intensity increase
3. Introduction of straight run with progressive speed increase
4. Proprioceptive training

21.2.3 Phase III (7 Days)

The transition criteria from the second to the third rehabilitation phase were clinical assessment and imaging examination (Bisciotti et al. 2019):

Clinical and functional criteria:

1. No pain during maximal concentric contractions
2. No pain during submaximal eccentric contractions

Imaging criteria

1. Disappearance of the lesion gap in MRI (or US) examination
2. Presence of compact granulation repair tissue within the lesion gap in US examination

Methods used during the third phase

1. Eccentric exercises with progressively increased intensity
2. Run with changes of direction at progressive speed
3. Football individual technique
4. Progressive integration with the team

21.3 Clinical Test for Return to Training

Concerning the return to training (RTT) decision-making process, in agreement with the Italian consensus statement on return to play after lower limb muscle injury in football (Bisciotti et al. 2019), the following requirements in clinical evaluation were met:

1. Absence of clinical symptoms.
2. Absence of pain or tenderness during muscle palpation.
3. Absence of pain on passive and active stretching.
4. Absence of pain on isometric, concentric, and eccentric contraction.
5. Completion of the prescribed rehabilitation program.
6. MRI and US imaging assessment.
7. Subjective feelings of the player were taken into account.

21.3.1 Specific Assessment

The specific assessment for short external hip rotator muscles injury was performed by means of the following tests (Delp et al. 1999; Busfield and Romero 2009; Delin et al. 2017):

1. Beatty maneuver test
2. Freiberg maneuver test
3. Internal rotation test

21.4 Laboratory Tests for RTT

In agreement with the Italian consensus statement on return to play after lower limb muscle injury in football (Bisciotti et al. 2019), the Pace and Nagle maneuvers assessed by dynamometric tests were performed.

21.5 Field Tests for RTT

In agreement with the Italian consensus statement on return to play after lower limb muscle injury in football (Bisciotti et al. 2019), the following tests for RTT decision-making process were performed:

1. Kicking test
2. Carioca test
3. Illinois Agility test
4. Agility *T*-test

21.6 Return to Play Tests

After obtaining clearance for RTT, in agreement with the Italian consensus statement on return to play after lower limb muscle injury in football (Bisciotti et al. 2019), the athlete started the return to play (RTP) decision-making process. The process was based on GPS data acquisition. During its 10-day duration, a certain numbers of parameters were recorded and divided in the following categories:

1. Quantitative evaluation
2. Qualitative evaluation
3. Parameter analysis

For an in-depth description of the parameters and categories, see Chap. 4.

After 10 days, the clearance for RTP was obtained. About 1 year after the injury, no reinjury occurred, and the subject is fully satisfied with the rehabilitation path.

21.7 Discussion

The IO is a flat and triangular muscle that is part of the hip extra-rotator muscles. It originates from the inner surface of the obturator membrane and the surrounding bony margins of the obturator foramen. Its muscular bands converge sideways and backward, then pass through the ischial foramen to become extra pelvic and give origin to a tendon that moves forward and beyond. Finally, the IO enters into the medial surface of the great trochanter, in front of and above the trochanteric pit. At the level of its distal insertion, the IO blends with the tendon of the gemellus inferior and superior muscles (all together called the triceps coxae) forming a common tendon (Yoo et al. 2015). This anatomical association with the gemellus inferior and superior tendon can explain the lower number of injuries that the IO shows compared to the external obturator muscle (Wong-On et al. 2018). The IO extra-rotates the thigh by abducting it when the latter is flexed, and it also contributes to the stabilization of the hip joint. It is innervated by the obturator nerve (L5, S2), and its blood supply depends upon the obturator artery. The IO maximum strength is related to the flexion degree of the hip joint and is maximal at the beginning of the oscillation phase during the run stride (Solomon et al. 2010). In addition to the IO, the other muscles working as hip external rotators are piriformis, external obturator, inferior gemellus, superior gemellus, and quadratus femoris. These muscles are known under the collective name of "short lateral rotators" (Gudena et al. 2015). This muscle group performs a major hip-joint stabilizing action, by stabilizing the femoral head within the acetabulum during hip movements (Yoo et al. 2015). Despite the important work done by this muscle group, little is found in literature. The existing studies mainly describe chronic syndromes, such as piriformis syndrome or ischial-femoral impingement (Meknas et al. 2009), and only few studies are based on acute injuries. In particular, to our knowledge, those dealing with acute injuries at IO level are only six, five of which are case reports (Bisciotti et al. 2019; Khodaee et al. 2015; Velleman et al. 2015; Caoimhe et al. 2016; Kelm et al. 2016), while one is an observational study (Wong-On et al. 2018). In their study Wong-On et al. (2018) reported the injury data recorded in four seasons in the professional Spanish football league (LIGA). During the considered period, the external obturator injuries recorded were 12, while the OI injuries were only four. However, it should be emphasized that the clinical diagnosis of IO injury presents some objective difficulties. Indeed IO-related lesions usually cause painful symptoms at the anterior hip-joint level. For this reason, IO injuries are in differential diagnosis with many other clinical frameworks causing GPS (Weir et al. 2015; Bisciotti et al. 2016; Griffin et al. 2016). The most reliable clinical test for IO injury is the passive internal rotation maneuver with and without muscle opposition by the patient (Kelm et al. 2016; Wong-On et al. 2018). Despite the fact that in general hip extra-rotatory muscle lesions, and those of the IO in particular, represent a rare occurrence (Kelm et al. 2016; Wong-On et al. 2018), these latter must necessarily be considered in clinical evaluation of a traumatic GPS framework.

The main mechanisms causing traumatic injury to the IO are essentially three:

1. The first is an unstable pelvis position with a sudden change in body weight distribution (Wong-On et al. 2018; Cass 2015).
2. The second is a sudden change of direction in condition of pelvic instability (Cass 2015; Wong-On et al. 2018).
3. The third, which represents the etiopathogenetic situation described in this study, is represented by a sudden hip intra-rotation with hip and knee flexed at about 90° (Delp et al. 1999).

The fact that reinjuries of the IO are not described in literature (Wong-On et al. 2018) confirms the hypothesis that they are caused by well-defined mechanical situations, like those described above, and are independent of predisposing intrinsic factors, such as excessive retraction or stiffness of the IO (Wong-On et al. 2018). The case presented in this study shows, in line with other studies, that the IO injuries present a good prognosis and a relatively short recovery time (Kelm et al. 2016; Wong-On et al. 2018).

21.8 Conclusions

The IO lesions are the rarest lesions among those involving the hip external rotators, which is a muscle group showing a very low incidence of lesion. Despite their low incidence, they must be considered in the case of a traumatic GPS framework. The IO injury etiopathogenesis is linked to well-defined mechanical situations and seems independent from intrinsic factors. In any case, IO injuries show a good prognosis and relatively short recovery times.

References

Bisciotti GN, Volpi P, Zini R, et al. Groin Pain Syndrome Italian Consensus Conference on terminology, clinical evaluation and imaging assessment in groin pain in athlete. BMJ Open Sport Exerc Med. 2016;2:e000142. https://doi.org/10.1136/bmjsem-2016-000142.

Bisciotti GN, Volpi P, Alberti G, Aprato A, Artina M, Auci A, et al. Italian consensus statement (2020) on return to play after lower limb muscle injury in football (soccer). BMJ Open Sport Exerc Med. 2019;5(1):e000505.

Busfield BT, Romero DM. Obturator internus strain in the hip of an adolescent athlete. Am J Orthop (Belle Mead NJ). 2009;38(11):588–9.

Caoimhe B, Alkhayat A, O'Neill P, Eustace S, Kavanagh E. Obturator internus muscle strains. Radiol Case Rep. 2016;12(1):130–2.

Cass SP. Piriformis syndrome: a cause of nondiscogenic sciatica. Curr Sports Med Rep. 2015;14(1):41–4.

Delin C, Vandensteen JY, Roger B. Hip short external rotator muscles injuries. In: Roger B, Guermazi A, Skaf A, editors. Muscle injuries in sport athletes. Cham: Springer; 2017.

Delp SL, Hess WE, Hungerford DS, Jones LC. Variation of rotation moment arms with hip flexion. J Biomech. 1999;32(5):493–501.

Griffin DR, Dickenson EJ, O'Donnell J, Agricola R, Awan T, Beck M, Clohisy JC, Dijkstra HP, Falvey E, Gimpel M, Hinman RS, Hölmich P, Kassarjian A, Martin HD, Martin R, Mather RC, Philippon MJ, Reiman MP, Takla A, Thorborg K, Walker S, Weir A, Bennell KL. The Warwick Agreement on femoroacetabular impingement syndrome (FAI syndrome): an international consensus statement. Br J Sports Med. 2016;50:1169–76.

Gudena R, Alzahrani A, Railton P, Powell J, Ganz R. The anatomy and function of the obturator externus. Hip Int. 2015;25(5):424–7.

Kelm J, Ludwig O, Schneider G, Hopp S. Injury of the obturator internus muscle- a rare differential diagnosis in a soccer player. Sportverletz Sportschaden. 2016;30(1):50–3.

Khodaee M, Jones D, Spittler J. Obturator internus and obturator externus strain in a high schoolquarterback. Asian J Sports Med. 2015;6(3):e23481. https://doi.org/10.5812/asjsm.23481.

Meknas K, Kartus J, Letto J, Christensen A, Johansen O. Surgical release of the internal obturator tendon for the treatment of retro-trochanteric pain syndrome: a prospective randomized study, with long-term follow-up. Knee Surg Sports Traumatol Arthrosc. 2009;17:1249–56.

Solomon LB, Lee YC, Callary SA, Beck M, Howie DW. Anatomy of piriformis, obturator internus and obturator externus: implications for the posterior surgical approach to the hip. J Bone Joint Surg Br. 2010;92(9):1317–24.

Velleman MD, Jansen Van Rensburg A, Janse Van Rensburg DC, Strauss O. Acute obturator internus muscle strain in a rugby player: a case report. J Sports Med Phys Fitness. 2015;55(12):1544–6.

Weir A, Brukner P, Delahunt E, Ekstrand J, Griffin D, Khan KM, Lovell G, Meyers WC, Muschaweck U, Orchard J, Paajanen H, Philippon M, Reboul G, Robinson P, Schache AG, Schilders E, Serner A, Silvers H, Thorborg K, Tyler T, Verrall G, Jan de Vos R, Vuckovic Z, Hölmich P. Doha agreement meeting on terminology and definitions in groin pain in athletes. Br J Sports Med. 2015;49:768–74.

Wong-On M, Turmo-Garuz A, Arriaza R, Gonzalez de Suso JM, Til-Perez L, Yanguas-Leite X, Diaz-Cueli D, Gasol-Santa X. Injuries of the obturator muscles in professional soccer players. Knee Surg Sports Traumatol Arthrosc. 2018;26(7):1936–42.

Yoo S, Dedova I, Pather N. An appraisal of the short lateral rotators of the hip joint. Clin Anat. 2015;28(6):800–12.

Case Report: Return to Play and Return to Training After Iliopsoas Injury

22

22.1 Introduction

This chapter is based on a case report describing an iliopsoas acute injury. At today, in the literature, the studies describing acute iliopsoas injuries are very rare. The limited number of studies describing iliopsoas acute injuries may be attributed to the objective difficulties in clinical diagnosis (Fredberg et al. 1995). Some authors reported that in a review of 4682 hip and pelvis injuries examined by means of MRI, 33 iliopsoas injuries were found, of which only 16 were found in adults younger than 65, and in this population only ten subjects were athletes (Bui et al. 2008). In that study, half of athlete's injuries were at muscle-tendon junction level, while the other half were partial tendon tears. Both in clinical and MRI examination, the iliopsoas injuries are often described in combination with adductor longus (15.9% of the cases), rectus femoris (8.5% of the cases), rectus abdominis (2.4% of the cases), and sartorius injuries (2.4% of the cases) (Serner et al. 2015). Furthermore, iliacus and psoas major injuries often occur together (Serner et al. 2018). A relatively common iliopsoas injury is the avulsion of the iliopsoas tendon with detachment of the lesser trochanter, secondary to an athletic injury. This type of injury is generally found in the growing skeleton before the lesser trochanter apophysis fusion (Bui et al. 2008). Complete iliopsoas tendon tears without lesser trochanteric avulsion are uncommon injuries (Theologis et al. 1997). Some studies described spontaneous rupture of the iliopsoas tendon in association with the lesser trochanter fracture (Lecouvet et al. 2005). In any case, acute iliopsoas muscle injuries and partial tendon tears are predominant in male athletic population and are most commonly athletics-related injuries (Bui et al. 2008). Indeed, Bui et al. (2008) reported that the patients under 65 years only showed muscle strains and partial tears and that the subjects with partial tears were, on average, almost 20 years older than those showing muscle strains. Furthermore, in the subjects over 65 years, the complete iliopsoas tendon tears were more common than either of the less severe injuries. Finally, the older the patients' population, the more severe the iliopsoas injury.

22.2 Case Report

A 28-year-old professional male soccer player came to our clinical evaluation complaining pain at right inguinal region. The pain had arisen 2 days earlier during a kicking movement performed during a training session. The pain sensation was present both during walking and when running and sometimes even in a static standing position. At the clinical examination the injured limb did not appear swollen, turgid, or painful to the palpation. There was no visible lesion gap on the muscle profile. The iliopsoas slightest passive elongation generated a severe painful symptomatology (VAS 7/10). The same pain was elicited

G. N. Bisciotti et al., *Return-to-Play after Lower Limb Muscle Injury in Football*,
https://doi.org/10.1007/978-3-030-84950-4_22

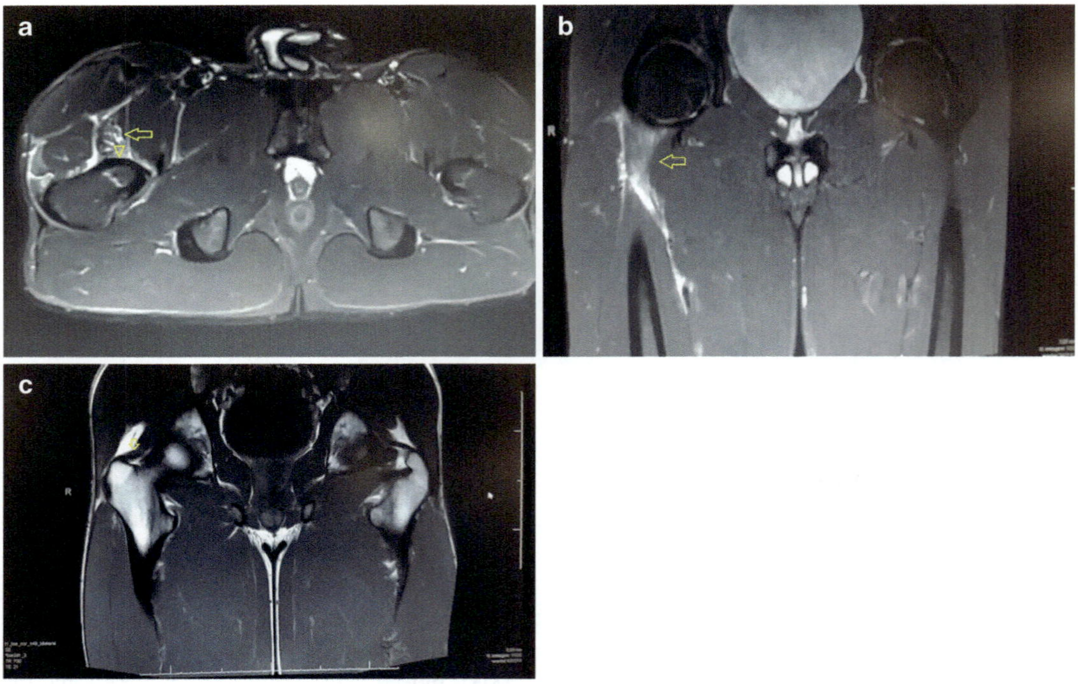

Fig. 22.1 Axial MRI STIR image (**a**) showing a first-degree lesion at iliopsoas level (arrow) close to the iliofemoral ligament (head arrow). (**b**) The same injury observed in coronal STIR (arrow). (**c**) in coronal T1 MRI the iliofemoral ligament appears intact (arrow)

during the iliopsoas isometric, concentric, and eccentric contraction (VAS 5/10, −6/10, /8/10, respectively). Therefore, on the basis of clinical tests, the diagnosis of iliopsoas indirect injury in its medial portion was made.

The MRI examination (Fig. 22.1) showed an iliopsoas indirect injury at the level in which the muscle passes near the iliofemoral ligament. The iliofemoral ligament was intact. The injury was classified as first degree following the classification proposed by the Italian Consensus Conference on guidelines for conservative treatment of lower limb muscle injuries in athletes (Bisciotti et al. 2019). The athlete started his rehabilitation program with the phase I.

22.2.1 Phase I (7 Days)

Methods used during the first phase

1. Pain control
2. Mobilization
3. Isometric exercises with progressive intensity increase
4. Hydrokinesitherapy

22.2.2 Phase II (7 Days)

The transition criteria from the first to the second rehabilitation phase were clinical assessment and imaging examination (Bisciotti et al. 2019):

Clinical and functional criteria:

1. Absence of pain during maximum isometric contraction
2. No pain during passive stretching
3. No pain during active stretching tests
4. Full ROM of the hip joint

Imaging criteria:

1. Decrease of the lesion gap at MRI (or US) examination
2. Presence of granulation repair tissue within the lesion gap in US examination

Methods used during the second phase

1. Active and passive stretching
2. Concentric exercises with progressive intensity increase
3. Introduction of straight run and spinning training with progressive speed increase
4. Proprioceptive training

22.2.3 Phase III (7 Days)

The transition criteria from the second to the third rehabilitation phase were clinical assessment and imaging examination (Bisciotti et al. 2019):

Clinical and functional criteria:

1. No pain during maximal concentric contractions
2. No pain during submaximal eccentric contractions

Imaging criteria:

1. Disappearance of the lesion gap in MRI examination
2. Presence of compact granulation repair tissue within the lesion gap in US examination

Methods used during the third phase

1. Eccentric exercises with progressively increasing intensity
2. Run with changes of direction at progressive speed
3. Individual football technique
4. Progressive integration with the team

22.3 Clinical Test for Return to Training

Concerning the return to training (RTT) decision-making process, in compliance with the Italian consensus statement on return to play after lower limb muscle injury in football (Bisciotti et al. 2019), the following requirements in clinical evaluation were met:

1. Absence of clinical symptoms.
2. Absence of pain or tenderness during muscle palpation.
3. Absence of pain on passive and active stretching.
4. Absence of pain on isometric, concentric, and eccentric contraction.
5. Completion of the prescribed rehabilitation program.
6. MRI and US imaging assessment.
7. Subjective feelings of the player were taken into account.

22.3.1 Specific Assessment

The specific assessment was based on the Thomas test (Harvey 1998; Ferber et al. 2010).

22.3.2 Laboratory Tests for RTT

In agreement with the Italian consensus statement on return to play after lower limb muscle injury in football (Bisciotti et al. 2019), the laboratory test for RTT performed was the Thomas test (Harvey 1998; Ferber et al. 2010), which was carried out in the following modalities:

1. In isometric contraction with the use of a dynamometer
2. In concentric contraction with the use of a dynamometer
3. In eccentric contraction with the use of a dynamometer

22.4 Field Tests for RTT

In agreement with the Italian consensus statement on return to play after lower limb muscle injury in football (Bisciotti et al. 2019), the following tests for RTT decision-making process were performed:

1. Kicking test
2. Illinois Agility test
3. Agility *T*-test

22.5 Return to Play Tests

After obtaining clearance for RTT, in agreement with the Italian consensus statement on return to play after lower limb muscle injury in football (Bisciotti et al. 2019), the athlete started the return to play (RTP) decision-making process. The RTP-DMP was based on GPS data acquisition. The process lasted 8 days, during which a certain number of parameters were recorded and divided in the following categories:

1. Quantitative evaluation
2. Qualitative evaluation
3. Parameter analysis

For an in-depth description of the parameters and categories, see Chap. 4.

After 10 days the clearance for RTP was obtained. About 8 months after the injury, no reinjury occurred, and the subject is fully satisfied with the rehabilitation path.

22.6 Discussion

This case report is focused on an iliopsoas isolated injury. Yet, as lesions of the iliopsoas often couple with other muscles injuries (in particular, adductor longus, rectus femoris, and rectus abdominis) both during clinical and imaging examination, it is very important to verify this possibility.

The iliopsoas musculotendinous unit injuries, although relatively uncommon, are more and more frequently detected because of the increase of athletic activity in the general population as well as the increasing demand to professional athletes. Iliopsoas tendon injuries are also encountered in the elderly population, for example, in the well-known complication of total hip arthroplasties (Chalmers et al. 2017). Most iliopsoas diseases, including tendinosis, avulsion injuries, snapping tendon, and bursitis, can present as groin pain syndrome (Bisciotti et al. 2016) and carries an extensive differential diagnosis. Furthermore, additional pathological abnormalities, such as an iliopectineal bursitis, may make the diagnosis more challenging. The imaging (US and MRI) examinations allows increased accuracy in the diagnosis of iliopsoas pathologies (Shabshin et al. 2005). Some authors (Tsukada et al. 2018) classified the changes in signal intensity in the iliopsoas pathologies, as visualized on short tau inversion recovery MRI examination, into two types:

1. The acute muscle injuries that is characterized by a massive high-signal area in the muscle belly, with a clear border
2. The peritendinitis, characterized by a long and thin high-signal area extending proximally along the iliopsoas tendon from the lesser trochanter, without a clear border

The authors concluded that the time needed for return to play was significantly shorter for athletes who had acute muscle injury changes in MRI signal intensity.

The iliopsoas injury clinical signs are quite uniform. The patient complaints pain at lower abdominal level in general lateral to the rectus abdominis muscle and above the inguinal ligament. In some patients, abdominal wall tenderness may also be observed, most frequently in the lower part of the rectus abdominis muscle. The pain is worsened with the effort during movements associated with the hip flexion and the motion of forward kicking of the ball (Mozes et al. 1985). In general, the pain is located at the iliopsoas muscle or its tendon level and could be provoked by pressure on this area. The most suggestive test for acute iliopsoas injury is the Thomas test (Harvey 1998; Ferber et al. 2010). The Thomas test must be performed in the supine position flexing the athlete leg maximally in the attempt to isolate the iliopsoas muscle from the other hip flexor muscles (Hölmich 2015).

Although it is a rather rare lesion, if correctly treated, the iliopsoas injury does not entail particular problems for the RTP in football. The basic point of the rehabilitation process is the respect of the biological stages of muscle repair. For this reason, the program we have adopted respected during the first, second, and third phase a very precise hierarchy of muscle contractions (i.e., isometric, concentric, and eccentric, respectively). This has ensured the maximum respect for the biology of the muscle tissue being repaired.

22.7 Conclusions

The anamnesis and physical examination are important for determining the iliopsoas injuries. However, MRI could play an important role in the diagnostic work-up. Indeed the MRI can help the clinician differentiate the isolated acute iliopsoas injuries from the iliopsoas "tandem lesion," that is, the iliopsoas injuries coupled with adductors longus, rectus femoris, or rectus abdominis injuries.

References

Bisciotti GN, Volpi P, Zini R, et al. Groin Pain Syndrome Italian Consensus Conference on terminology, clinical evaluation and imaging assessment in groin pain in athlete. BMJ Open Sport Exerc Med. 2016;2:e000142. https://doi.org/10.1136/bmjsem-2016-000142.

Bisciotti GN, Volpi P, Alberti G, Aprato A, Artina M, Auci A, et al. Italian consensus statement (2020) on return to play after lower limb muscle injury in football (soccer). BMJ Open Sport Exerc Med. 2019;5(1):e000505.

Bui KL, Ilaslan H, Recht M, Sundaram M. Iliopsoas injury: an MRI study of patterns and prevalence correlated with clinical findings. Skelet Radiol. 2008;37(3):245–9.

Chalmers BP, Sculco PK, Sierra RJ, Trousdale RT, Berry DJ. Iliopsoas impingement after primary total hip arthroplasty: operative and nonoperative treatment outcomes. J Bone Joint Surg Am. 2017;99(7):557–64.

Ferber R, Kendall KD, McElroy L. Normative and critical criteria for iliotibial band and iliopsoas muscle flexibility. J Athl Train. 2010;45(4):344–8.

Fredberg U, Hansen LB, Kissmeyer-Nielsen P, Torntoft EB. Iliopsoas tendonitis in athletes. Diagnosis and treatment. Ugeskr Laeger. 1995;157:4031–3.

Harvey D. Assessment of the flexibility of elite athletes using the modified Thomas test. Br J Sports Med. 1998;32(1):68–70.

Hölmich P. Groin injuries in athletes--development of clinical entities, treatment, and prevention. Dan Med J. 2015;62(12):B5184.

Lecouvet FE, Demondion X, Leemrijse T, Vande Berg BC, Devogelaer JP, Malghem J. Spontaneous rupture of the distal iliopsoas tendon: clinical and imaging findings, with anatomic correlations. Eur Radiol. 2005;15:2341–6.

Mozes M, Papa MZ, Zweig A, Horoszowski H, Adar R. Iliopsoas injury in soccer players. Br J Sports Med. 1985;19(3):168–70.

Serner A, Tol JL, Jomaah N, Weir A, Whiteley R, Thorborg K, Robinson M, Hölmich P. Diagnosis of acute groin injuries: a prospective study of 110 athletes. Am J Sports Med. 2015;43(8):1857–64.

Serner A, Weir A, Tol JL, Thorborg K, Roemer F, Guermazi A, Yamashiro E, Hölmich P. Characteristics of acute groin injuries in the hip flexor muscles - a detailed MRI study in athletes. Scand J Med Sci Sports. 2018;28(2):677–85.

Shabshin N, Rosenberg ZS, Conrado FA. MR imaging of iliopsoas musculotendinous injuries. Magn Reson Imaging Clin N Am. 2005;13(4):705–16.

Theologis TN, Epps H, Latz K, Cole WG. Isolated fractures of the lesser trochanter in children. Injury. 1997;28:363–4.

Tsukada S, Niga S, Nihei T, Imamura S, Saito M, et al. Iliopsoas disorder in athletes with groin pain. Prevalence in 638 consecutive patients assessed with MRI and clinical results in 134 patients with signal intensity changes in the iliopsoas. JB JS Open Access. 2018;3(1):e0049.

GPSR Compliance

The European Union's (EU) General Product Safety Regulation (GPSR) is a set of rules that requires consumer products to be safe and our obligations to ensure this.

If you have any concerns about our products, you can contact us on ProductSafety@springernature.com

In case Publisher is established outside the EU, the EU authorized representative is:

Springer Nature Customer Service Center GmbH
Europaplatz 3
69115 Heidelberg, Germany

Batch number: 10732801

Printed by Printforce, the Netherlands